VETERINARY CLINICAL PROCEDURES
IN SMALL ANIMAL PRACTICE

VETERINARY CLINICAL PROCEDURES

IN SMALL ANIMAL PRACTICE

VICKI JUDAH, AS, CVT

CENGAGE
Learning®

Australia • Brazil • Mexico • Singapore • United Kingdom • United States

Veterinary Clinical Procedures in Small Animal Practice
Vicki Judah

Vice President, Careers & Computing: Dawn Gerrain

Product Manager: Dan Johnson

Director, Development-Career and Computing: Marah Bellegarde

Product Development Manager: Juliet Steiner

Senior Content Developer: Darcy M. Scelsi

Editorial Assistant: Andrew Oiumet

Market Development Manager: Scott Chrysler

Senior Production Director: Wendy Troeger

Senior Content Project Manager: Elizabeth Hough

Senior Art Director: David Arsenault

Production Service: Prashant Kumar Das, MPS Limited

Compositor: MPS Limited

Cover image(s): ©Julija Sapic/Shutterstock.com

For product information and technology assistance, contact us at
Cengage Learning Customer & Sales Support, 1-800-354-9706
For permission to use material from this text or product,
submit all requests online at **www.cengage.com/permissions**
Further permissions questions can be e-mailed to
permissionrequest@cengage.com

Library of Congress Control Number: 2014939096
ISBN: 978-1-4354-6962-4

Cengage Learning
200 First Stamford Place, 4th Floor
Stamford, CT 06902
USA

Cengage Learning is a leading provider of customized learning solutions with office locations around the globe, including Singapore, the United Kingdom, Australia, Mexico, Brazil, and Japan. Locate your local office at:
www.cengage.com/global

Cengage Learning products are represented in Canada by Nelson Education, Ltd.

To learn more about Cengage Learning, visit **www.cengage.com**

Purchase any of our products at your local college store or at our preferred online store
www.cengagebrain.com

Notice to the Reader

Publisher does not warrant or guarantee any of the products described herein or perform any independent analysis in connection with any of the product information contained herein. Publisher does not assume, and expressly disclaims, any obligation to obtain and include information other than that provided to it by the manufacturer. The reader is expressly warned to consider and adopt all safety precautions that might be indicated by the activities described herein and to avoid all potential hazards. By following the instructions contained herein, the reader willingly assumes all risks in connection with such instructions. The publisher makes no representations or warranties of any kind, including but not limited to, the warranties of fitness for particular purpose or merchantability, nor are any such representations implied with respect to the material set forth herein, and the publisher takes no responsibility with respect to such material. The publisher shall not be liable for any special, consequential, or exemplary damages resulting, in whole or part, from the readers' use of, or reliance upon, this material.

Printed in the United States of America
Print number: 01   Print Year: 2014

To all the students and countless others who dedicate their lives to the care and understanding of our companion animals.

CONTENTS

CHAPTER 2   RABBITS   123

CHAPTER 3 FERRETS 169

CHAPTER 4 GUINEA PIGS 207

CHAPTER 5   SMALL RODENTS   241

CHAPTER 6   BIRDS   281

CHAPTER 7   REPTILES   335

PREFACE

Many skills and techniques must be mastered in becoming a veterinary technician—skills that apply to most small animal species, yet can be adaptable to safely and successfully perform procedures for the different species. Learning new skills is progressive, and range from restraint techniques or a basic examination to blood draws, radiography, and anesthesia. With a sound basis, each species included in this book begins by building and expanding on the previously learned techniques.

The format of *Veterinary Clinical Procedures in Small Animal Practice* follows the successful approach to learning presented in the *Veterinary Clinical Procedures in Large Animal Practice*. The chapters and procedures are not written in paragraphs of prose but, rather, present the material so the student may easily understand not only the purpose for the procedure but also the equipment needed, step-by-step instructions, and potential problems that may be encountered. In addition, each step is amplified and explained with short statements and comments, a rationale that provides not just the *how*, but just as important, the *why* of a procedure.

Because of the ever growing, ever changing profile of a small animal practice, techniques for exotic species are also included. Clients expect the veterinary staff to be familiar with more than just their companion animals and also competent and knowledgeable about the care of their other nontraditional pets.

STUDENT RESOURCES

You may also like:
> *An Illustrated Guide to Veterinary Medical Terminology*, 4e, by Janet Romich
> > ISBN: 978-1-1331-2576-1
> *The Veterinary Technician's Pocket Partner – A Quick Access Reference Guide*, by Marisa Bauer
> > ISBN: 978-1-4283-5782-2
> *Restraint and Handling for Veterinary Technicians and Assistants*, by Bonnie Ballard and Jody Rockett
> > ISBN: 978-1-4354-5358-6

INSTRUCTOR RESOURCES

Spend less time planning and more time teaching with Cengage Learning's Instructor Resources to Accompany *Veterinary Clinical Procedures in Small Animal Practice*. This content can be accessed through your Instructor SSO account.

To set-up your account:

- Go to **www.cengagebrain.com/login**
- Choose **Create a New Faculty Account**.
- Next you will need to select your **Institution**.
- Complete your personal **Account Information**.
- Accept the **License Agreement**.
- Choose **Register**.
- Your account will be pending validation; you will receive an e-mail notification when the validation process is complete.
- If you are unable to find your Institution; complete an **Account Request Form**.

Once your account is set up or if you already have an account:

- Go to www.cengagebrain.com/login
- Enter your e-mail address and password and select **Sign In**.
- Search for your book by author, title, or ISBN.
- Select the book and click **Continue**.
- You will receive a list of available resources for the title you selected.
- Choose the resources you would like and click **Add to My Bookshelf**.

Components available on the Instructor Companion site include the following:

INSTRUCTOR'S MANUAL

An electronic Instructor's Manual provides instructors with invaluable tools for preparing for class lectures and examinations.

COMPUTERIZED TEST BANK

An electronic testbank makes and generates tests and quizzes in an instant, with a variety of types of question, including multiple-choice, fill-in-the-blank, and matching. This testbank consists of questions that test students thoroughly on retention and application of what they've learned in the course. Answers are provided for all questions so instructors can focus on teaching, not grading.

INSTRUCTOR POWERPOINT SLIDES

A comprehensive offering of instructor support slides created in Microsoft® Power Point outlines concepts and objectives to assist instructors with lectures.

ACKNOWLEDGMENTS

The author would like to thank the following persons for their generous time and willingness to offer help and advice and recognize that, without them, this book and my experience in writing it would have been far less rich:

Kathy Nuttall for providing many of the new photographs and technical advice; Ben Beck for his assistance in merging files; Tegan Spangrude; Carmen Val Dez and her very cooperative iguana, " Bubba Fat"; Beverlee Brown, who without a pause offered to print, fax, and scan "anything, anytime"; Craig Ashby for " holding the reins"; and Dr. Eric Klaphake, DVM ("quick question, but an important one"); and, "Solo," my little Perro de Baja for being so good about being a demonstrator-dog for photos and techniques, and his patience with it all. Good boy!

REVIEWERS

Anne Duffy, AS, BA, MA
Kirkwood Community College
Cedar Rapids, Iowa

Sarah Lefebvre, DVM
Mount Ida College
Newton, Massachusetts

Lois Sargent, DVM
Miami Dade College
Miami, Florida

Lewanne E. Hunt Sharp, BA, RVT, VTS (ECC)
Mesa Community College
Mesa, Arizona

DOGS AND CATS

The greatness of a nation and its moral progress can be judged by the way its animals are treated.

—Gandhi

UNIT ONE

Overview of Species

OBJECTIVES

Upon completion of this unit, the reader should be able to:

▶ Identify species characteristics

▶ Identify and appropriately address animal behavior

KEY TERMS

piloerection

cephalic vein

lateral saphenous vein

DOGS

Dogs are the most popular companion animals and dogs of all breeds are seen daily in a busy veterinary practice (**Figure 1-1**). Patients may present for a variety of reasons varying from a new puppy health examination to a major trauma or a critical illness. There are many breeds of purebred dogs and multiple others that are of mixed breeding, some for which the parentage is known and others of an unknown combination of breeds. Regardless of breed, their care, handling, and medical concerns are, with few exceptions, the same. Stray dogs may be brought in either by a member of the public or by animal services, and these dogs may not be well socialized and present a greater risk to the veterinary staff members who handle them.

▲ **FIGURE 1-1** Dogs are the most popular companion animals presented to the veterinary practice.

BEHAVIOR

A fearful dog is much more likely to bite and with little provocation. Some signs of fear include: dilated pupils, a cowering crouched body posture, and avoidance of eye contact (**Figure 1-2**). The tail is tucked up along the belly or, in short-tailed dogs, clamped down tight; it may back itself into a corner, shaking, and quivering with its hackles raised. The dog may yelp or whine, urinate, or defecate. Great care must be taken when approaching a fearful dog. Avoid direct eye contact; speak in a soft, reassuring voice using the dog's name; and do not approach from the front but, instead, turn the side of your body toward the dog. Do *not* lean forward over the top of the dog or attempt to pat the top of the dog's head or neck.

▲ **FIGURE 1-2** It is important to recognize a fearful dog's body language.

The aggressive dog must be approached with care and handled safely to avoid injury. Signs of aggression in dogs include a forward-facing posture, ears erect with the tail held straight out or curled up and over the back (**Figure 1-3**). **Piloerection** is evident; the hairs on the back of the neck and on the rump area are elevated. The lips may be curled in a snarl to expose the teeth. The dog will growl deeply, bark aggressively, and may make an initial lunge. *A slowly wagging tail is not a sign that the dog is relaxing and becoming friendly but is another warning that the dog is very likely to bite.* All behaviors must be carefully observed and recognized before attempting to handle an aggressive dog, regardless of its size.

▲ **FIGURE 1-3** An aggressive dog exhibits all the signs and movement that an attack is imminent and should never be further provoked.

CATS

Cats are second only to dogs in popularity (**Figure 1-4**). Like dogs, there are many breeds of purebred cats; however, the majority of cats seen in a veterinary practice are of mixed breeding and are generally referred to as domestic cats. Domestic cats are categorized as either domestic short hair (DSH) or domestic long hair (DLH). Examples of purebred breeds are Siamese, Persian, and Manx. Cat fanciers also may refer to their cats by their distinctive markings—for example, the "Tuxedo Cat" which is not a breed, but is so called because of the black coat and white markings that resemble a tuxedo.

▲ **FIGURE 1-4** Cats are second to dogs as companion animals.

BEHAVIOR

A cat with aggressive/defensive behaviors will growl, hiss, and "scream" loudly. A hiss may be followed by a "popping" noise, a sign of serious intent. The ears will be turned back and the eyes narrowed and focused (**Figure 1-5**). A cat demonstrating these behaviors is likely to attack if it cannot escape, and great care must be taken to avoid personal injury.

▲ **FIGURE 1-5** This cat showing defensive/aggressive body language. Great care must be taken with both aggressive cats and fearful cats (Figure 1-6). In both situations the cat is likely to bite and scratch in an attempt to escape.

Fearful cats sometimes attempt to hide and make themselves appear as small as possible in the farthest corner available, with their tails and all four paws tucked close-in under the body. With some cats, the fur, especially the tail hairs, are completely erected (piloerection) in an attempt to make themselves appear larger (**Figure 1-6**). Their pupils are dilated, and they may hiss and growl.

▲ **FIGURE 1-6** A cat showing fearful body language.

The majority of dogs and cats are beloved family members and must be treated with kindness, compassion, and knowledge. This extends not just to the patients, but also their human companions, the clients who entrust their care to veterinary staff.

UNIT TWO

Restraint Techniques

OBJECTIVES

Upon completion of this unit, the reader should be able to:

▶ Explain different methods of restraint

▶ List the reasons for restraint

▶ Identify the equipment used to facilitate restraint

▶ Demonstrate various restraint techniques used with the dog and cat

KEY TERMS

anal sacs

auscultate

cat bag

catchpole

chemical restraint

medial saphenous vein

muzzles

physical restraint

sedative

slip leash

sternal recumbency

tranquilizer

GUIDELINES FOR RESTRAINT

The number-one cause of injury in a veterinary practice results from animal bites. It is important to understand that all patients can and will bite. Some breeds or types are more aggressive than others, but many bites occur because the animal is afraid or because it is ill or in pain. Every patient will react differently to a given situation, regardless of breed or breed characteristics. The skilled veterinary technician will not only recognize breeds and general characteristics of their temperaments but also be skilled in interpreting body language and behavior in any given situation. It is also important for the veterinary professional to understand that a situation and the animal's reaction to it can change dramatically within seconds. All of the attitudes of the presenting patient must be understood and interpreted carefully prior to attempting any restraint technique.

SIGNS OF PAIN IN ANIMALS

It is important not only to recognize and interpret signs of aggression and fear, but also the signs of an animal in pain:

- Vocalizing when touched or moved (growling, hissing, whining, yelping)
- Restlessness (pacing, circling, getting up and down frequently)
- Reluctance to move or to lie down
- Avoidance
- Piloerection
- Hunched or "tucked up" posture
- Limping or holding a limb abnormally
- Chewing at the site or perceived source of pain

METHODS OF RESTRAINT

Mastering restraint techniques is essential for every veterinary technician. Correct methods of restraint are necessary for the successful and safe completion of every procedure, from a basic physical examination to veinpuncture, radiographic positioning, and anesthesia induction.

Two methods of restraint are **physical restraint** and **chemical restraint**. Physical restraint means that the animal is *physically* held by a person in such a manner that it cannot escape or cause harm to itself or to personnel. Often, in physical restraint, various tools (equipment) are used to help to control the animal. Chemical restraint involves the use of a **sedative** or a **tranquilizer** and, in some instances, general anesthesia when the patient is injured and attempted physical restraint may cause further complications or pain.

Sedatives differ from tranquilizers. Sedatives usually are given to reduce irritability and excitement. Sedatives depress the central nervous system and, depending upon the dose, may induce a state of complete relaxation, drowsiness, or even sleep. Tranquilizers also calm a fractious patient, but they are generally regarded as less potent and may, in some circumstances, increase the excitability of the patient.

The manner in which a patient is restrained depends on the temperament of the animal, the extent of pain involved, the length of the procedure, and most important, the reason for restraint. Even when restraint is performed by the most experienced of hands and potential complications are minimized, restraint can affect some animals adversely.

RESTRAINT PROCEDURES

Dogs, in general, are socialized "family members" and are accustomed to being handled gently, with petting, grooming, and affection. A well-socialized, friendly dog is far easier to deal with than a dog that has been poorly trained, abused, even verbally, or has had little or no interaction with people.

Cats typically are not as socialized as dogs and may already be frightened by the car ride, the strange surroundings, odors, and the presence of other cats and dogs. It is a good idea to allow the cat to settle in its own carrier before attempting any type of restraint.

Physical restraint, however brief, can be disturbing to an animal. *Always know the reason for restraint and the method required for the procedure before approaching the patient.*

Using your voice is important, to reassure the patient or to make a recognized command to the dog, such as "Sit." The tone and pitch of your voice are also important: Most dogs respond better to higher-pitched voices and "silliness," words that sound like fun. Commands should be stated clearly and firmly. Cats rarely respond to training commands, but it is just as important to use a soft and gentle voice when reassuring them. Different procedures require different restraint techniques, but remember: *Always use the minimum amount of restraint required to perform the procedure.*

PET OWNERS AND RESTRAINT

However well intended, owners should not be allowed to assist in the restraint or attempt to reassure their own animals while they are being restrained. If owners are injured or bitten by their own animals during the course of a veterinary procedure, the veterinary practice is held liable.

PURPOSE

- To safely and securely control the patient for a variety of procedures
- To move the patient from one area of the facility to another
- To provide for the safety of the patient and the staff

COMPLICATIONS

- Trauma including bruises and lacerations
- Fracture of long bones and spinal injury
- Asphyxia and strangulation
- Evacuation of bladder, bowel, and **anal sacs**
- Exacerbating existing injuries
- Injury to personnel

EQUIPMENT

- Nylon or rope slip leash
- Muzzles
- Length of rolled gauze
- Restraint pole
- Heavy towels/blanket
- Cat bag

LEATHER GLOVES

Leather gloves are rarely a benefit when attempting to handle fractious or aggressive patients. Because of their bulk and size, it is impossible to obtain a secure hold on an animal while wearing gloves. Although they do offer some protection for the handler, they should not be used in an attempt to restrain a patient for a procedure, but only to facilitate another of method of restraint

PROCEDURE for Using the Slip Leash

The most commonly used piece of equipment is a simple nylon or rope **slip leash** with a ring on one end and a hand loop on the other (**Figure 1-7**). It is lightweight and easily slipped over the animal's head. It can be used when removing a patient from a kennel or cage or to restrain an aggressive animal behind a kennel door (**Figure 1-8.**) It should never be used to tie a patient to a fixture because of the danger of choking.

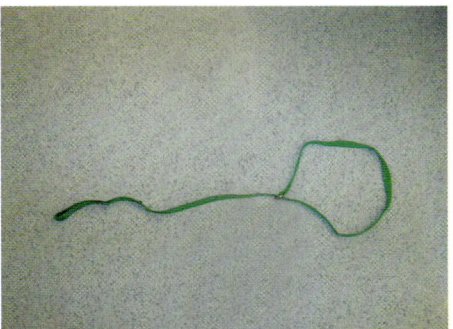

▲ **FIGURE 1-7** A slip leash, one of the most simple and frequently used items for restraint.

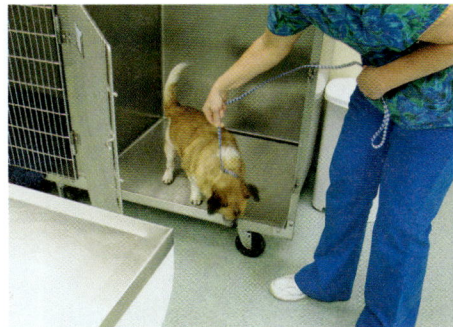

▲ **FIGURE 1-8** A slip leash being used to take a dog out of the cage.

TECHNICAL ACTION	RATIONALE/AMPLIFICATION
1. Put the loop handle through the ring at the other end, forming a closed circle.	
2. Hold the leash in one hand and quickly slip it over the animal's head.	
3. Pull the noose snug around the animal's neck.	3a. The noose should not be pulled so tight that it chokes the patient.

(Continues)

TECHNICAL ACTION	RATIONALE/AMPLIFICATION
4. Put your hand through the loop handle so it is around your wrist, and take hold of the leash below so you are not holding the loop end in your hand. (**Figure 1-9**)	**4a.** This prevents a large dog from jerking the leash from your hand (**Figure 1-9**). Slip leashes are also used to restrain a cat in a cage, and to draw it forward where a more suitable method of restraint can be used.

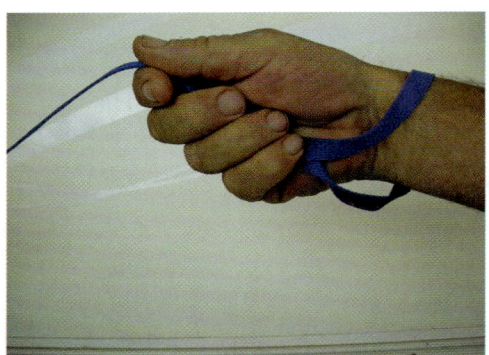

▲ **FIGURE 1-9** The correct method to hold a slip leash. Placing the loop around the wrist provides a more secure hold and helps prevent it from being pulled out of the hand of the restrainer.

PROCEDURE for Use of a Muzzle

Muzzles are used to prevent an animal from opening its mouth and biting. Dog muzzles are made of leather with a buckle end or nylon with a quick-release buckle (**Figure 1-10**). Both types are adjustable. The benefit of nylon muzzles is that they are easily washed and disinfected and, because of this, they are the most commonly used type of muzzle.

© terekhov igor/Shutterstock.com

▲ **FIGURE 1-10** An example of a leather dog muzzle. The correct size and type of muzzle should always be used for handler safety and to prevent pressure on the dog's muzzle and area below the eyes.

Cat muzzles are also made of nylon, designed to cover the entire head except the nostrils, and have Velcro® fastenings (**Figure 1-11**).

With experience, muzzles are easily and quickly placed on a dog or a cat. Care must be taken not to cover the nostrils or to tighten the muzzle so much that it puts any

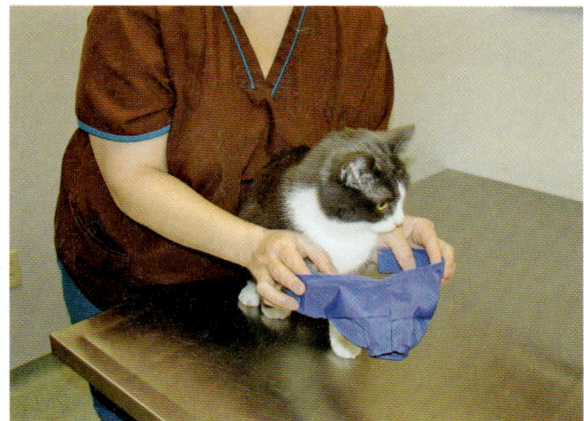

▲ **FIGURE 1-11** An example of a nylon cat muzzle. The technician is in the correct position for muzzle placement.

pressure under the patient's eyes. This is seen most often when the muzzle is too large and the muzzle is pulled too tightly in an attempt to "make it fit." Physical restraint is also required, as dogs are adept at using their forelegs and claws and can quickly remove a muzzle.

Unlike dog muzzles, cat muzzles are designed to cover the entire head and eyes, with an opening at the end to leave the nose uncovered. Covering cats' eyes has a calming effect; if they cannot see, they cannot react as quickly. Physical restraint is also required to avoid injury from their claws and to keep the patients safe.

TECHNICAL ACTION	RATIONALE/AMPLIFICATION
1. Choose the correct size muzzle and make sure the fastening in the back is not broken and the straps are secure.	
2. Approach the dog from behind and quickly place the muzzle over the dog's mouth (**Figure 1-12**).	2a. This is to avoid being directly in front of the dog (confronting it) and the possibility of being bitten.

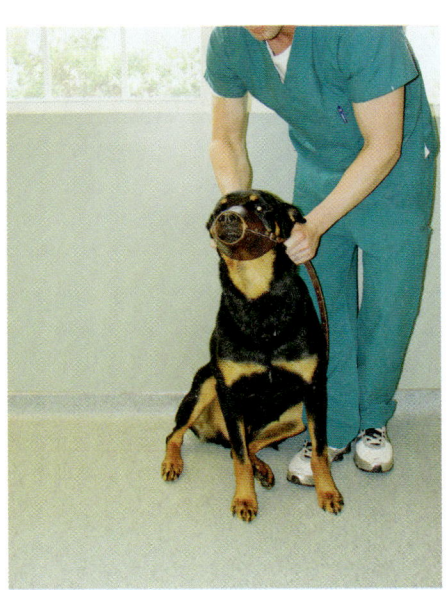

◀ **FIGURE 1-12** Approach the dog from behind and place the muzzle over the dog's mouth.

(Continues)

TECHNICAL ACTION	RATIONALE/AMPLIFICATION

3. Secure the fastening at the back of the head, just below the ears, and adjust the fit (**Figure 1-13**).

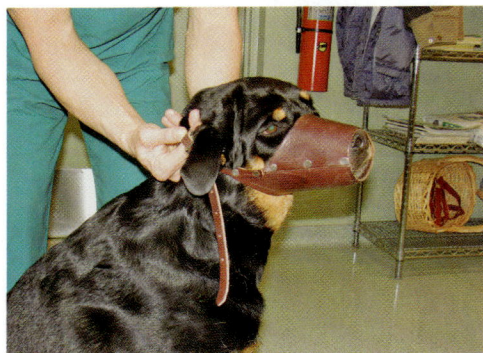

▲ **FIGURE 1-13** Secure the muzzle behind the dog's ears, making sure that it isn't too tight or that it puts pressure on the eyes.

3a. Dog muzzles should not put any pressure on the eyes.

3b. Cat muzzles must always be positioned so that the opening in the front encircles the nares and allows the cat to breathe (**Figure 1-14**).

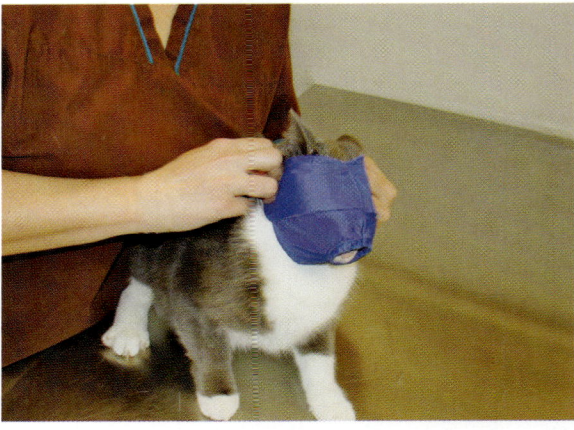

▲ **FIGURE 1-14** When applying the cat muzzle, be sure the opening is over the nares so the cat is able to breathe.

4. Apply cat muzzles from behind the head.

USE OF A SLIP LEASH OR GAUZE FOR A MUZZLE

A length of gauze or a slip leash also may be used to form a secure muzzle. Gauze is easier to apply if it is wet, which gives it more substance. Make a loop in the gauze or leash, slip it over the dog's muzzle, and quickly pull the ends under the jaws to tighten it. Tie the ends under the dog's mouth, then up behind the dog's head, and secure the ends in a double bow (**Figures 1-15a** and **1-15b**).

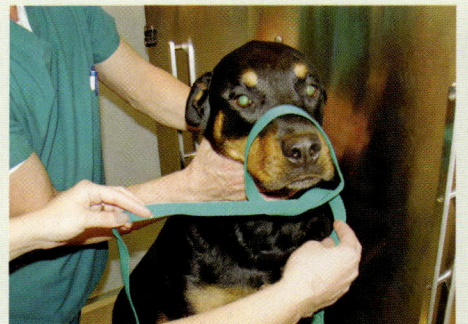

▲ **FIGURE 1-15a** A slip leash is easily converted to a dog muzzle. Make a loop with the leash and place it over the dog's nose with a single wrap of both ends under the jaw.

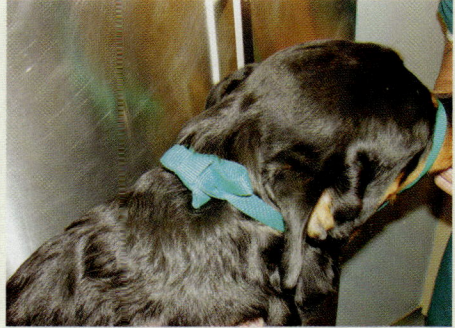

▲ **FIGURE 1-15b** Loop both ends and tie it securely behind the dog's head with a quick release bow.

PROCEDURE for Using a Catchpole

The **catchpole**, sometimes referred to as a "rabies pole," is used to capture or snare a loose animal that is otherwise unapproachable (**Figure 1-16**). It is alsois used to control an aggressive patient while keeping it at a safe distance from the handler (**Figure 1-17**). Be prepared for the animals to immediately react to the tightening of the snare. They sometimes struggle violently and attempt to twist out of the snare. *Use with caution:* If misused, this device is capable of strangling the animal. It should never be used to lift an animal off the ground or to hoist it into a cage by its neck, but only to control the animal until a tranquilizer can be administered and become effective. With an extremely difficult animal, two catchpoles can be used, with the handlers on opposite sides, keeping the animal centered at a safe distance between the two poles and handlers.

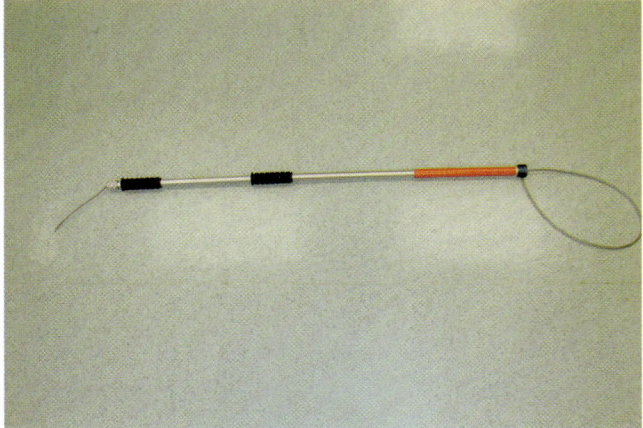

▲ **FIGURE 1-16** An example of a catchpole with a wire loop on the end.

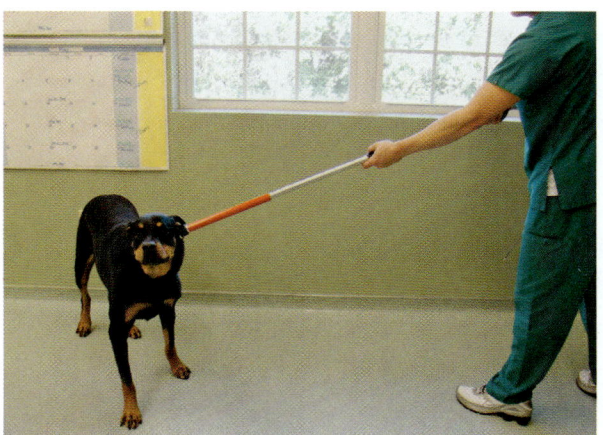

▲ **FIGURE 1-17** Holding the catchpole correctly to maintain a safe distance from the animal.

TECHNICAL ACTION	RATIONALE/AMPLIFICATION
1. Become very familiar with the catchpole, especially the quick-release mechanism, before attempting to use it.	1a 1b. Refer to Figures 1-16 and 1-17.
2. Extend the wire noose so that it settles over the animal's head and quickly pull the handle to tighten the snare.	
3. Firmly hold the catchpole with both hands to keep the animal away at the maximum distance.	

PROCEDURE for Using the Cat Bag

The **cat bag** is used to prevent the cat from clawing the restrainer and from escaping. It will not prevent the cat from biting. The bag is made of heavy nylon with zippered openings to withdraw a single limb. It is used to administer injections or to perform venipuncture.

TECHNICAL ACTION	RATIONALE/AMPLIFICATION
1. Check to be sure that all four of the side zippers are fully closed.	
2. Place the cat bag on the table with the top zipper fully open (**Figure 1-18**).	
3. Lift (or scruff) the cat, and in one smooth, rapid maneuver place the cat inside the bag and zip the top opening shut (**Figure 1-19**).	**3a.** Take care not to catch any fur/skin when zipping the top closed.
4. Fasten the Velcro® strap around the cat's neck (**Figure 1-20**).	
5. Hold the cat bag with both hands around the shoulders, behind the head to prevent the possibility of a bite.	**5a.** A fractious cat may "bounce" the bag off the table onto the floor, causing injury and increased panic at being "trapped."
6. Open one of the side zippers to allow access to withdraw a limb to perform the required procedure (**Figure 1-21**).	

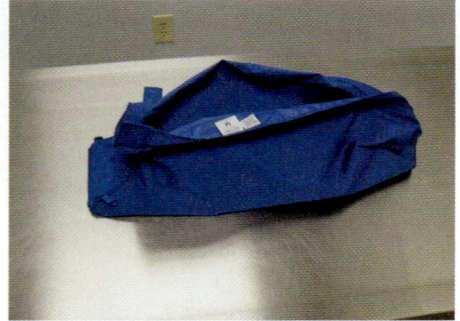

▲ **FIGURE 1-18** An example of a nylon cat bag.

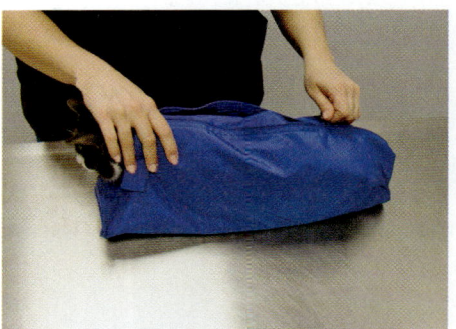

▲ **FIGURE 1-19** Place the cat in the bag and zip the bag up to the cat's neck.

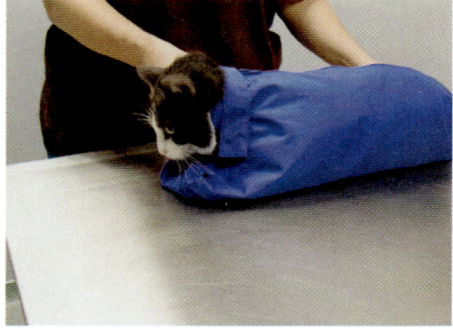

▲ **FIGURE 1-20** Secure the bag around the cat's neck with the Velcro® fastening.

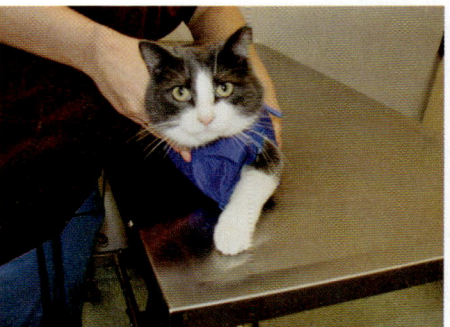

▲ **FIGURE 1-21** By opening a zipper on the side of the bag, one limb can be withdrawn from the bag to perform a blood draw, give an IM injection in a rear leg, or perform nail trims.

Many cats can be restrained with the use of towels. With reassurance and a soft voice, the cat can be placed in the center of a large towel. The towel is wrapped around the body of the cat, and an assistant can gently hold the cat in the towel and restrain the head if necessary.

Some cats can be restrained with just a light scruff, and the restrainer can distract the cat, if necessary, by patting *very* gently on the flat of the nose. (*Never* flick a cat—or any other animal—on the nose for any reason. It is not only cruel, but unnecessary, unreasonable, and very painful to the patient.) Scruffs should be minimal both in time and the pressure or grip applied. Always use the minimum amount of restraint necessary to perform the procedure, and this is especially important to remember when handling feline patients.

PROCEDURE for Standing Restraint

The standing restraint is used during a physical examination, to take a patient's temperature, or when expressing anal sacs.

TECHNICAL ACTION	RATIONALE/AMPLIFICATION
1. Place a slip leash around the patient's neck.	1a. Always use a slip leash. Never attempt to restrain the dog using its own collar, as the dog may pull out of it.
2. Position yourself so that you are facing the person performing the procedure.	2a. You are, in effect, presenting the patient to the person performing the procedure who should have unobstructed access.
3. Use one arm to restrain the head, and put the other arm under the abdomen to restrain the body (**Figure 1-22**).	3a. Always turn the patient's face away from the person performing the procedure to reduce the chance of being bitten.
	3b. By holding the dog or cat under the abdomen, you prevent it from sitting down.

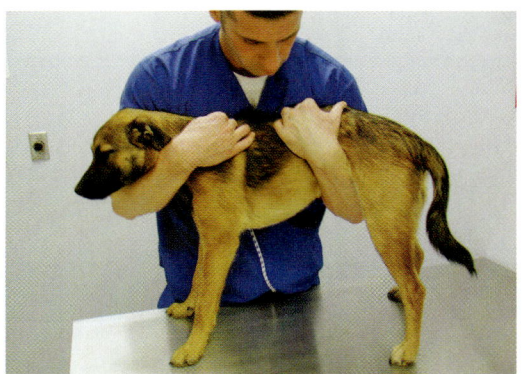

▲ **FIGURE 1-22** Restraint of the dog in a standing position for a basic examination.

4. Hold the patient close to your own body, and adjust your hold as the procedure requires.	4a. Holding the patient close to your body blocks the patient from moving away from the person performing the procedure. Your body becomes a barrier and also offers some comfort to the patient.

(Continues)

TECHNICAL ACTION	RATIONALE/AMPLIFICATION

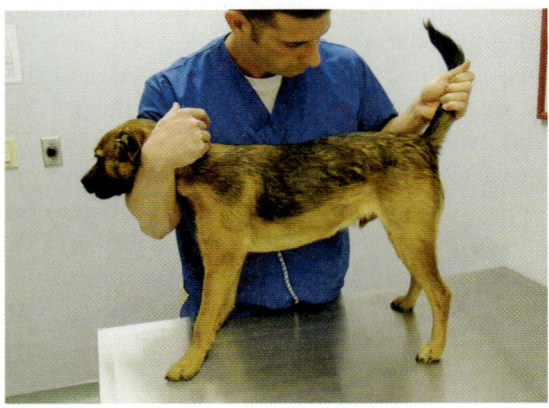

▲ **FIGURE 1-23** Standing restraint holding the tail up. This method facilitates obtaining a rectal temperature or expressing anal sacs.

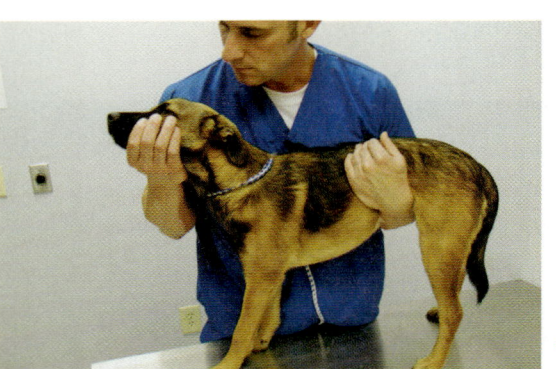

◄ **FIGURE 1-24** Standing restraint with the restrainer's arm re-positioned to give clear access for auscultation of the heart and lungs.

4b. Minor adjustments to this method of restraint can be made by holding the tail erect. This may be necessary for taking a rectal temperature, expressing anal sacs, or obtaining a fecal sample (**Figure 1-23**). This restraint also may be modified so the examiner can auscultate (listen) for heart and chest sounds (**Figure 1-24**).

4c. Large dogs may resist being placed on the examination table more than the actual restraint. It is often easier to restrain them and perform the procedure while they are on the floor by kneeling down next them.

PROCEDURE for Restraint of a Sitting Dog

A patient is frequently restrained in a sitting position for examination of the head, eyes, and ears. It may also be used in presenting a foreleg for a cephalic blood draw.

TECHNICAL ACTION	RATIONALE/AMPLIFICATION

1. Place the slip leash around the dog's neck.

2. Tell the dog to "Sit" (remember—your voice is also a restraint technique).

3. The dog may sit obediently on command or you may have to gently press down on the hindquarters, and say "Sit" again.

3a. Be careful. For a dog with arthritis or other joint problems, this may be painful. Allow the patient to sit in the most comfortable position, which may be with the weight more on one hip or the other. It doesn't always have to be a "square-bottom sit."

(Continues)

TECHNICAL ACTION	RATIONALE/AMPLIFICATION
4. With the dog in a sitting position, place one arm around the dog's neck so the head is turned slightly toward you, and *always* away from the person performing the procedure. Place the other arm under the dog's chest (**Figure 1-25**).	4a. The ear of the dog may held or stroked while restraining the head, but it should never be twisted in an attempt to gain more control. Twisting of the pinna can cause permanent damage to the cartilage of the ear, and causes the dog unnecessary pain and or discomfort. It may also be interpreted by the dog as aggression, as fighting dogs often grab the ears of their opponent (refer to Figure 1-25.)

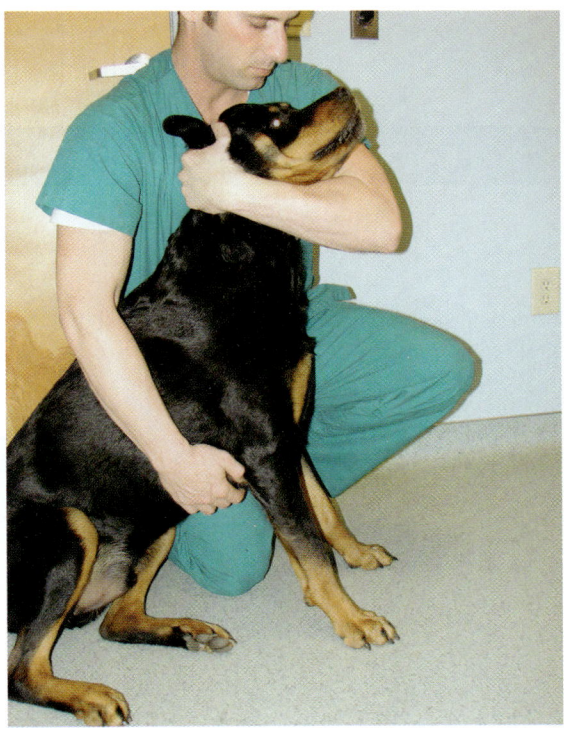

◀ **FIGURE 1-25** Restraint of the dog in a sitting position. The ear may be held gently as ear rubs are often a comfort to the dog. The ear should never be pinched, twisted, or used in an attempt to achieve greater control during physical restraint.

RESTRAINT of the Cat in a Sitting Position

Cats are also restrained in a sitting position to examine the head, eyes, and ears.

TECHNICAL ACTION

1. Cats are not likely to sit on command.

2. Hold the cat's head with both hands cupped under the cat's jaws with the thumbs placed on top of the head (**Figure 1-26**).

3. Gently push the cat down into a sitting position with your elbow.

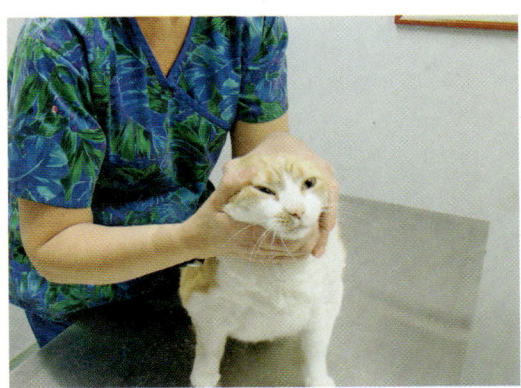

▲ **FIGURE 1-26** Restraint of the cat in a sitting position.

PROCEDURE for Restraint of the Dog in Sternal Recumbency

Another restraint technique is to place the patient in **sternal recumbency**. This positions the patient in the down position, resting on its chest (sternum). The head, eyes, ears, and mouth can be easily examined. This is also a good restraint technique for presenting the foreleg for a **cephalic** blood draw without the need to place the patient in lateral recumbency, or flat on its side. The cephalic vein is located on the foreleg.

TECHNICAL ACTION	RATIONALE/AMPLIFICATION
1. Place a slip leash around the dog's neck.	
2. Restrain the patient in a sitting position.	
3. Place one arm around the dog's neck and the other arm across its back, and grasp the forelimbs (**Figure 1-27**).	3a. If the dog is reluctant to lie down, gently pull the forelimbs forward.

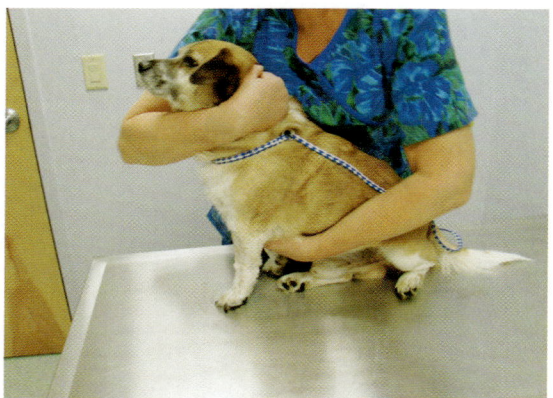

▲ **FIGURE 1-27** To place a dog in sternal recumbency, begin with a sitting restraint.

4. Use your voice, "Down," while using your body to push the dog down onto its chest.

5. Lean over the dog and hold it close to your body (**Figure 1-28**).

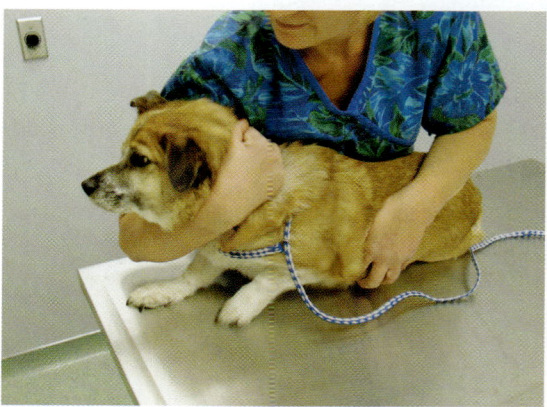

▲ **FIGURE 1-28** Restraint of the dog in sternal recumbency.

PROCEDURE for Restraining the Cat in Sternal Recumbency

Cats, like dogs, are often placed in sternal recumbency to examine the eyes, head, and mouth, as well as to present a foreleg for a cephalic blood draw. Cats are often more relaxed in sternal recumbency than in a sitting position. Always provide a soft towel or blanket for cats to rest on, as this is more relaxing and less frightening for them than a cold table.

TECHNICAL ACTION	RATIONALE/AMPLIFICATION
1. Put a towel on the examination table and place the cat on the towel.	1a. Cats often settle themselves into sternal recumbency once placed on the table and there may be no need for steps 2, 3, and 4.
	1b. With cats, always use a soft voice. Cats also relax with touch and gentle scratching under the chin.
	1c. A good distraction is to sprinkle loose catnip flakes or a catnip spray onto the towel. Catnip has a soporific effect and calms many a nervous or frightened cat.
2. Gently push the cat down into a sitting position.	
3. When the cat is sitting, put a little pressure on the shoulders with your other hand.	
4. When the cat is on its chest, use both arms to hold the cat in position and further restrain it by holding the head (**Figure 1-29**).	

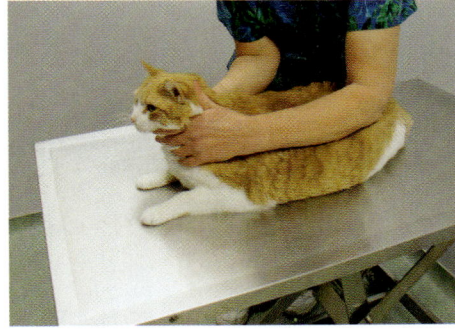

▲ **FIGURE 1-29** Restraint of the cat in sternal recumbency.

PROCEDURE for Restraining the Dog in Lateral Recumbency

In lateral recumbency, the dog is laid on one side of the body. This restraint technique is often used to have more control over an uncooperative patient and to examine the underbelly, abdomen, and genitals for wounds and abnormalities. In this position, the cephalic vein (located along the medial (inner) foreleg) and the **lateral saphenous** vein (located on the lateral (outside) of a rear leg) may also be presented for a blood draw.

TECHNICAL ACTION	RATIONALE/AMPLIFICATION
1. Place a slip leash around the dog's neck.	
2. Begin the restraint as you would for sternal recumbency.	

(Continues)

TECHNICAL ACTION	RATIONALE/AMPLIFICATION

3. Place your right arm across the dog's neck and grasp the patient's right foreleg, using the other hand to grasp the right rear limb (**Figure 1-30**).

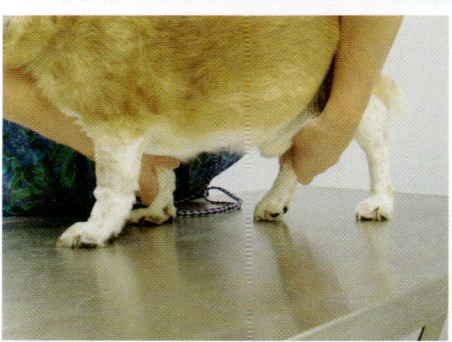

▲ **FIGURE 1-30** Correct positioning of the hands to restrain a small dog in lateral recumbency.

4. Slightly lift both limbs, and gently roll the patient onto its side (**Figure 1-31**).

4a. Do this gently but swiftly. Never slam the patient onto the table.

5. Maintain the restraint by placing your right arm across the dog's neck, and hold the forelegs in the right hand; use your left hand to restrain the rear legs. Reverse the positions of your arms if required or you are more comfortable and confident as the restrainer.

5a. If the dog is large and struggles when placed on its side, brace your forearm on the patient's neck and use your hands to restrain the down-side legs. This prevents the dog from getting up (refer to Figure 1-31).

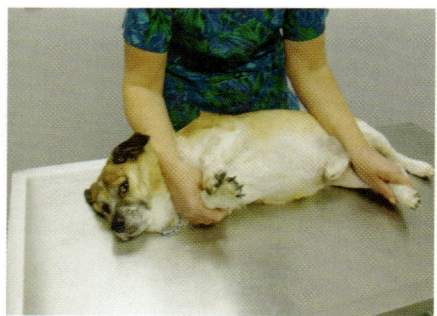

▲ **FIGURE 1-31** Restraint of a dog in lateral recumbency.

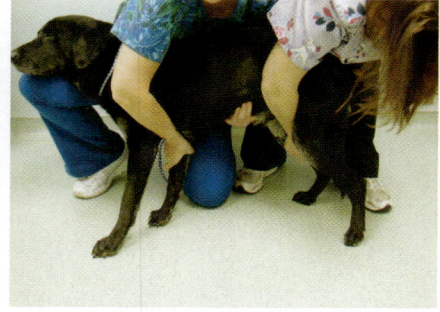

▲ **FIGURE 1-32** Two restrainers are sometimes required for a large, uncooperative patient. One person controls the head and forelegs, and the other person controls the hind quarters and rear limbs. Once the patient has been rolled onto its side, both restrainers will be required to hold the patient in position. By holding the downside limbs, the patient will be unable to stand.

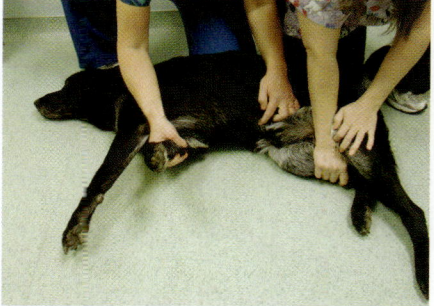

▲ **FIGURE 1-33** The restrainer of the hind limbs is also in position to present the upper rear leg for a lateral saphenous blood draw.

PROCEDURE for Restraint of the Cat in Lateral Recumbency

Cats are placed in lateral recumbency to further examine areas of concern on the underbelly, abdomen, and genitals. This method of restraint for the cat is often used to access the **medial saphenous** vein for a blood draw. Cats are not particularly comfortable in this position and are likely to struggle. It may help to place the towel over the cat's head.

TECHNICAL ACTION	RATIONALE/AMPLIFICATION

1. Place the cat on the table.

2. Scruff the cat and lift it off the table only enough to grasp both hind legs with your free hand.

3. Maintain the scruff and lay the cat gently on its side, extending the hind limbs (**Figure 1-34**).

    3a. This is also known as the "cat stretch." With this method, the forelegs are not restrained. (Refer to Figure 1-34)

    3b. The towel can be adjusted to cover the forelegs and offer some protection from potential scratches.

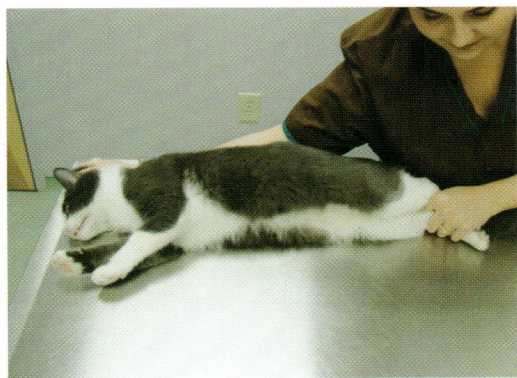

▲ **FIGURE 1-34** Restraint of the cat in lateral recumbency. This also referred to as the "cat stretch," as it is most often used to restrain uncooperative patients for further examination of the abdomen and genitals

RESTRAINT PROCEDURES FOR VENIPUNCTURE

One of the most common reasons for restraint is to obtain a blood sample. Before attempting restraint, it is important to communicate with the person performing the procedure and determine the vein of choice. Different venipuncture sites require different restraint techniques. The location of the most commonly accessed veins and the position required for restraint are included in the procedures for obtaining a blood sample.

UNIT THREE

Grooming

OBJECTIVES

Upon completion of this unit, the reader should be able to:

▶ Demonstrate basic grooming skills: bathing, nail trims, expressing anal sacs, and ear cleaning.

KEY TERMS

ataxia	fistula
axilla	mats
cerumen	otitis externa
cerumenolytic agent	otoscope
clipper burn	pinna
dewclaw	tympanic membrane
dips	

GUIDELINES FOR GROOMING

Patients are frequently presented for basic grooming, not necessarily for the skills required of a professional groomer but, rather, those of a medical nature or as part of a treatment plan. For example, a dog may have tangled with a skunk, be infested with fleas or other skin parasites, or require treatment with a medicated shampoo to treat a skin condition. Many times, dogs are presented for nail trims or to have their anal sacs expressed. Sometimes, hospitalized patients have to be bathed or they have a specific area that needs "cleaning up" either from becoming cage-soiled with urine and fecal material or to clean blood and debris from injuries. Frequently, dogs are presented for ear cleaning and aural treatments.

Cats may need to have their nails trimmed, ears cleaned, areas of matted hair shaved out (particularly in long-haired cats), and underlying skin conditions treated. They may be infested with fleas or ear mites or, because of age or illness, are no longer grooming themselves.

GROOMING PROCEDURES

The veterinary technician should be competent in performing all of the basic grooming tasks and become aware of conditions that may warrant further investigation. Often, during what appears to be a routine bath or ear cleaning, other problems are noted and should be directed to the attention of the veterinarian.

PURPOSE OF BATHING

- To improve the overall wellness of an animal
- To maintain or achieve a healthy coat and skin
- To treat for external parasites
- To treat skin disorders
- To cleanse soiled areas of a hospitalized patient and help prevent pressure sores
- To present a clean and well-groomed (cared for) patient upon discharge

PURPOSE OF NAIL TRIM

- To prevent overly long or ingrown nails
- To allow the animal to walk normally
- To help prevent traumatic fractures of the nail
- To reduce damage to household furnishings
- To minimize scratches to people and other animals

PURPOSE OF ANAL SAC EXPRESSION

- To relieve and remove the buildup of anal sac secretions
- To prevent discomfort, "scooting," irritation, and inflammation

PURPOSE OF EAR CLEANING

- To remove dirt, matted hair, and other debris from the ear
- To remove **cerumen** (ear wax) and foreign objects
- To treat **otitis externa,** an inflammation of the outer ear
- To clean prior to administering medications
- To clean after obtaining culture specimens

COMPLICATIONS OF BATHING

- Hypersensitivity or allergic reaction to shampoos or insecticidal dip
- Difficult or reluctant (hard to restrain) patient
- Risk of injury, strangulation, and death if left unattended
- Soap and solutions getting into the eyes and ears
- Water too hot
- Overheating/burning with the drier
- Clipper burn

COMPLICATIONS OF NAIL TRIM

- Cutting too close to the quick, causing bleeding and pain

COMPLICATIONS OF ANAL SAC EXPRESSION

- Severely impacted anal sacs that cannot be expressed manually
- Rupture of the anal sac
- Abscess of the anal sac

COMPLICATIONS OF EAR CLEANING

- Injury to the external ear canal
- Rupture of the **tympanic membrane** (ear drum)

EQUIPMENT FOR BATHING

- Assortment of brushes and combs
- Clippers with a #10 blade attached
- Cotton balls
- Protective eye-lubricant
- Shampoo and coat condition formulated for use in dogs
- Medicated shampoo (as prescribed)
- Insecticidal dip solution (as prescribed)
- Disposable gloves
- Plastic apron
- Slip leash
- Cloth or sponge
- Stack of clean, dry towels
- Standing blow dryer or cage dryer
- Tub with spray nozzle
- Warm water
- Sterile saline

EQUIPMENT FOR NAIL TRIM

- Appropriate size and type of nail trimmer
- Cauterizing agent (silver nitrate sticks or Quick Stop Powder)

EQUIPMENT USED FOR ANAL SAC EXPRESSION

- Disposable gloves
- Sterile lubricant

- Gauze pads
- Basin of warm soapy water to clean rectal area
- Warm water to rinse

EQUIPMENT USED FOR EAR CLEANING

- **Otoscope** (a hand-held instrument with a light source used to examine the inner ear)
- Basin
- Bulb syringe
- Cotton balls
- Cotton swabs
- Hemostats
- **Ceruminolytic agent** (a solution that breaks up cerumen (ear wax) so it is removed more easily)
- Sterile saline

PROCEDURE for Bathing a Dog

As stated earlier, a dog may need bathing for a variety of reasons: as a treatment for fleas or a skin disorder, to clean up blood or sites of injuries, or to clean urine or fecal material from the animal.

TECHNICAL ACTION	RATIONALE/AMPLIFICATION
1. Place a slip leash around the dog's neck.	1a. Always place the slip leash on the dog before removing its own collar. This enables better control and avoids wetting the dog's own collar, which may be expensive leather or a nylon collar that could hold soap residue, remain wet, and potentially cause an irritation.
	1b. A major reason for client displeasure is the staff losing or misplacing the dog's own collar and leash. Always label the collar. Many facilities maintain a row of hooks for hanging identified collars so they are not misplaced.
2. With the dog in standing restraint, thoroughly brush out the coat and remove all clumps of matted hair using the electric clippers if necessary.	2a. Brushing and combing out the coat removes dead hair and debris. Mats, clumps of tightly tangled hairs, frequently adhere to the skin and cause sores. Mats should be shaved out (*never* attempt to use scissors) using an electric clipper with a #10 blade attached. Clippers should be held flat against the skin and moved so the blade cuts under the mat. Clippers can quickly heat up, and the blade should be checked frequently to avoid causing **clipper burn**, when the blade becomes hot enough to actually burn the surface of the skin. Test the heat of the blade by holding it against your own arm. Either replace the blade or use one of the commercial aerosol sprays formulated to remove the heat from the blade, provide lubrication, and reduce friction.

(Continues)

TECHNICAL ACTION	RATIONALE/AMPLIFICATION

3. Place large cotton balls in each ear, being careful not to push the cotton ball into the ear canal.

4. Place a small amount of sterile eye ointment in each eye (**Figure 1-35**).

3a. This helps prevent water and soap from entering the ear canal, potentially causing an infection.

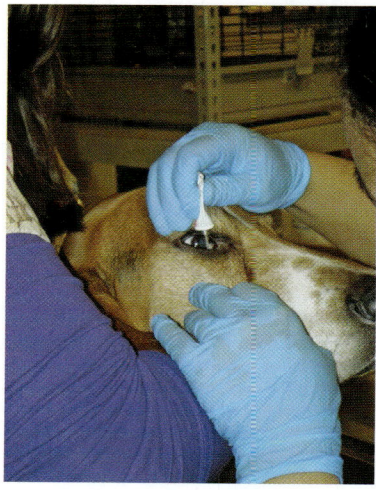

▲ **FIGURE 1-35** Prior to bathing, a sterile eye ointment should be applied to protect the eyes from cleaning products while bathing.

5. Don disposable gloves and plastic apron.

6. Lift the dog into the tub (tie or hold in place).

7. Slowly turn on the water and adjust the temperature.

7a. A sudden spray of water may startle the dog. The water should be warm, but never hot. Dogs have been known to suddenly collapse when the water becomes too hot. Check the temperature frequently.

8. Gradually introduce the water and thoroughly wet the coat, starting from just behind the head and moving toward the tail (**Figure 1-36**).

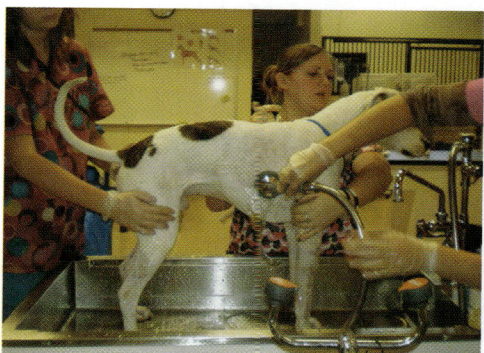

▲ **FIGURE 1-36** With the water nozzle, wet the patient prior to applying shampoo. Thoroughly rinse soap residue off the dog. Carefully check areas of the groin, axilla, underbelly, rectum, and tail.

9. Spread a line of shampoo from shoulders to rump.

10. Begin by massaging in the shampoo, adding water as necessary to create a sudsy foam.

11. Pour shampoo into the palm of your hand and apply to the abdomen, genitals, and rectal areas.

(Continues)

TECHNICAL ACTION	RATIONALE/AMPLIFICATION
12. Rinse hands and apply shampoo from your hands to the tail and each leg.	
13. Carefully lift each foot and clean between the toes and pads of the feet.	
14. Wet the dog's face with the cloth or sponge, add a little shampoo, and use it to clean the face and head.	
15. Starting at the back of the head, carefully and completely rinse all areas of the dog's body with warm water, paying particular attention to the underbelly, **axilla** (armpit), genitals, rectal area, between the toes, and pads of the feet.	15a. To avoid getting a good soaking yourself, wear a plastic apron and leave the head until last. The dog's natural reaction is to shake the water off its head quickly, followed by a whole-body shake. Using a wet cloth or sponge reduces the urge to shake, and this method also helps prevent accidentally getting soap in the dog's eyes.
16. Use a clean, wet, soap-free cloth or sponge to remove all soap from the dog's face and head.	
17. Remove as much water as possible, using your hands to wipe the water down and away.	
18. Use some of the towels to dry off as much water as you can before lifting the dog from the tub.	18a. Never allow the dog to jump from the tub. The floor is likely to be wet and slippery and could cause a bad fall for both you and the dog.
19. Remove the cotton balls from the dog's ears.	
20. Wrap the dog in a dry towel, and remove from the tub.	
21. Place the dog in a kennel with a cage dryer (**Figure 1-37**), a free-standing dryer. With small dogs, a hand-held blow drier also could be used.	21a. Monitor dogs carefully when using a cage drier. Turn the control to low heat to avoid overheating the dog. Do not put any towels on the floor of the cage. They only become wet and inhibit drying.

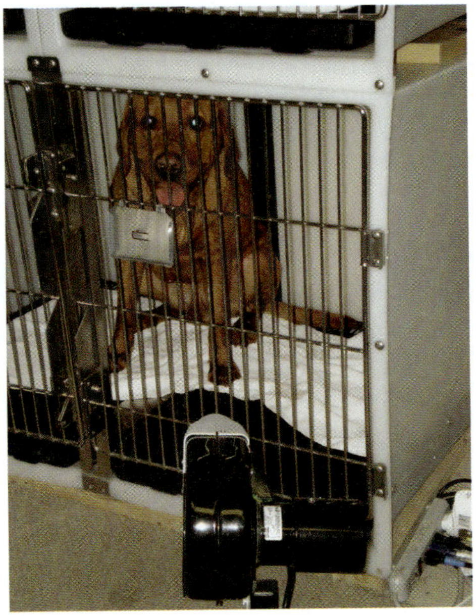

▲ **FIGURE 1-37** When using a dryer, monitor the temperature closely to avoid overheating the animal.

SKUNK ODOR

Although there are a few commercial products formulated to remove skunk odor, one of the most effective treatments is plain tomato juice. Thoroughly wet the dog's coat, and pour the tomato juice liberally all over the body, carefully avoiding getting the juice into the eyes or ears. (You may need a couple of gallons.) Wear an apron and disposable gloves and have a clean set of scrubs ready for yourself. Massage in the tomato juice and allow it to remain on the coat for approximately 15 minutes. Apply the tomato juice to the head and face with a sponge.

Most dogs, rarely cats, "get skunked" by trailing along behind the skunk, oblivious to the consequences or warning signs, and get sprayed in the face and eyes. Pay particular attention to the dog's eyes, and consult with the veterinarian regarding a prescribed, soothing eye ointment. After the elapsed amount of time, rinse the dog thoroughly. If the odor is still strong, repeat the tomato bath soak. There likely will be some residual smell, so follow up with a regular deodorizing shampoo and a cream rinse.

Another simple remedy is to mix one quart of hydrogen peroxide with one-quarter cup of baking soda, and add two teaspoons of a mild liquid soap. (For a large dog, double the ingredients). Pour this mixture over the dog and massage it into the coat. Let it stand for 5 to 15 minutes. The dog must be thoroughly rinsed off, then bathed with a deodorizing shampoo and conditioner. A warning with this method: Leaving it on the coat longer will not make it more effective, but it will change the color of the coat because of the hydrogen peroxide bleaching the hair coat. In addition, repeated applications can potentially damage the hair, making it brittle.

These are both treatment suggestions that you can offer to a client who telephones asking for help. It avoids bringing the dog into a clinic and having the skunk smell permeate the entire building.

PROCEDURE for Bathing a Cat

Cats, with a few exceptions, do not enjoy being bathed and can become difficult to control during the procedure. Cats usually don't become aggressive. Their only intent is escape, so it is very important to keep them under control and safe from escape and or injury.

TECHNICAL ACTION	RATIONALE/AMPLIFICATION
1. Make sure that all doors and windows are securely closed.	1a. Cats are more likely to attempt escape than a dog. Even if the cat gets away from you, it should not be able to get out of the room through an open door or window.
2. Be aware and prepared knowing that most cats do not like water or having a bath.	
3. As with the dog, have all supplies ready before you begin.	
4. Put a drain plug in the tub, and add warm water no more than 6 inches deep.	4a. Cats do not like the spray hose and are handled more easily if they are placed directly into the water.

(Continues)

TECHNICAL ACTION	RATIONALE/AMPLIFICATION
5. Prior to placing the cat in the tub, fit a feline harness and light leash to the harness. Calmly place the cat in the tub of water and reassure the cat with your voice.	5a. Normally, a harness and short leash is all that is required to control the patient for a bath. If the cat is struggling for escape, the use of a scruff may also be necessary. The use of a slip leash can be even more alarming to the cat and should not be used. Should the cat attempt to escape, a wet nylon slip leash can quickly become a strangling noose.
6. While maintaining a short leash or hand on the harness, pour the warm water gently over the cat's back, thoroughly wetting the fur. Use a cloth or sponge for the face. At this point, be prepared to also scruff the cat if it starts to struggle in an attempt to escape.	6a. Never use a glass jug, which could easily slip out of your hands and break.
7. Add the shampoo and massage in, using the same methods as for the dog.	7a. *It is extremely important to use only products that are approved for use in cats, as many products approved for use in dogs are toxic to cats.*
8. When finished shampooing, pull the plug and allow the water to drain from the tub. Refill it to the same depth with clear water to rinse the cat, using the basin. Repeat until no shampoo remains.	
9. When all residue of soap has been rinsed clean, drain the water and use your hands to squeeze as much water out of the fur as you can.	
10. Wrap a towel around the cat and return it to a cage to air dry or use a small hand-held hair drier on low heat.	

MEDICATED SHAMPOOS

There are many different types of medicated shampoos formulated to treat different types of skin problems and are prescribed by the veterinarian. All directions on the label must be carefully read, understood, and adhered to. All medicated shampoos require contact time—that is, leaving the shampoo on for a specified amount of time. Do not assume that shampoos are all alike in their mode of use and effectiveness. Many medicated shampoos that are used for dogs are toxic to cats. *Never* use a medicated shampoo on a cat unless it is specifically labeled that it is safe for use with cats.

Always wear disposable gloves, protective eye goggles, and an apron to protect yourself. People can also be allergic to some medicated shampoo products.

DIPS

Dips are insecticidal solutions that are poured over the coat of a clean, just bathed, and still damp patient. All precautions and warning labels on the product should be read, understood, and adhered to, especially with regard to the required dilution. Not only cats may have an adverse reaction, but some breeds of dogs are also known to have toxic reactions. Visible signs of an adverse reaction and toxicity include hyper-salivation (drooling), vomiting, and **ataxia** (inability to ambulate or move normally) and a severe reaction could cause seizures.

Any sign of change or abnormality in the patient requires immediate medical attention. Patients should always be observed for at least 20 minutes after they have been returned to the cage or kennel to ensure that no problem develops. Patients that have received a dip should be allowed to air-dry. Always brush out the coat when the patient is completely dry. Brushes and combs should be thoroughly cleaned after use to remove all chemical residues.

PROCEDURE for Nail Trims

Sometimes a patient's nails may become overgrown and may cause soreness or irritation. Keeping an animal's nails trimmed can prevent this.

TECHNICAL ACTION	RATIONALE/AMPLIFICATION

1. Place slip leash around the patient's neck.

2. Examine the toes and nail growth.

   Select appropriate type and size of nail trimmers (**Figure 1-38**). Have silver nitrate sticks or powdered Quik-stop® on hand, opened and ready for use. (**Figure 1-39**).

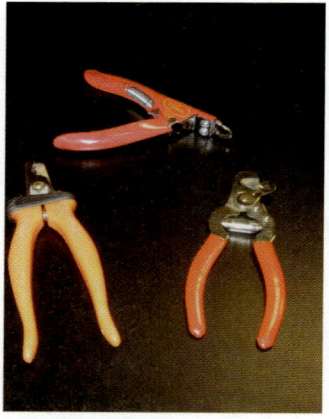

▲ **FIGURE 1-38** Several varieties and sizes of nail trimmers are available. Technicians should become familiar with the manner in which they are held and used and choose the appropriate type and size.

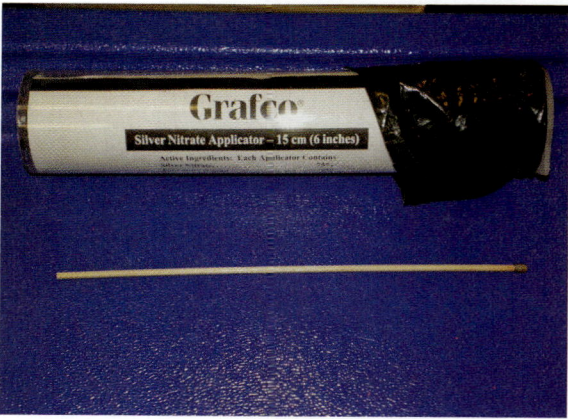

▲ **FIGURE 1-39** Silver nitrate sticks look similar to long match sticks. The chemical is on the very tip. Always place a used silver nitrate stick on a paper towel, not directly on a counter top as it will permanently stain the surface.

3. An additional restrainer may be necessary. The patient may be restrained while sitting or be placed in either sternal or lateral recumbency, depending on the temperament of the patient.

4a. Some dogs may have been "quicked" in the past. Quicking occurs when the nail is cut too short, causing pain and bleeding. Other dogs dislike and resist having their feet handled. Some dogs may have to be muzzled. It is always easier to ask for help with restraint than to struggle with a difficult patient on your own.

4b. Reluctant or fractious cats may be placed in a cat bag, using the zipper openings to access each foot. With some cats, only a towel wrap may be necessary, extending one leg at a time.

(Continues)

TECHNICAL ACTION	RATIONALE/AMPLIFICATION
4. Firmly grasp the foot and hold the toe between your thumb and forefinger.	
5. Push the toe forward to extend the nail.	6a. Cat claws are retractable. Gently push down on the toe, just caudal to the first joint, to extend the claw.
6. Determine where to clip, avoiding cutting into the quick. Refer to **Figure 1-40**.	

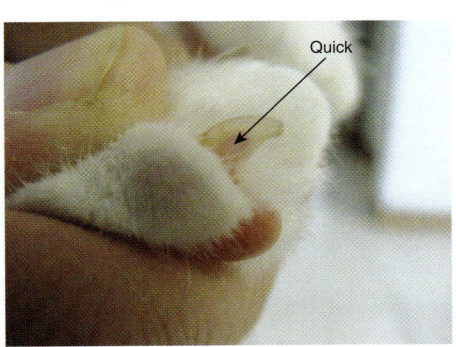

Quick

 FIGURE 1-40 The quick is clearly visible in the clear nail of a cat. The quick, or nail bed, follows the length of the nail. With dark or black nails, it is better to take smaller amounts of the nail to establish the location of the quick and avoid one quick cut that could injure the nail bed and cause bleeding.

7. Clip with one clean cut to prevent frayed or fractured edges.	8a. Do not forget the **dewclaw**, which is on the medial side of the foreleg in both dogs and cats. Many dogs have had the dewclaws removed. Some dogs also have dewclaws on the rear legs. These must all be trimmed to prevent an overgrowth that can sometimes curl around and penetrate the skin.
8. Examine the clipped nail for bleeding, and treat with a cauterizing agent before moving to the next toe.	9a. If using silver nitrate sticks, always place them on a thickness of paper towels or dispose of immediately. Once used, they will leave indelible dark marks on counters, clothes, and hands.

✳ THE DREMEL TOOL

An electric dremel type tool is frequently used not only to trim the nails but also to smooth away rough edges. If used for trimming the nails, it should be applied to the tip until the correct length is achieved. Dremel type tools also cauterize, and great care has be taken not to use this power tool to extremes, or as a "short cut" to standard nail trim procedures. Nails can easily be cut too short with a rotary powered tool and cause unnecessary pain. Deliberately cutting down to the quick is cruel and inappropriate. It may cause temporary lameness in the dog, and it most certainly will cause the dog to resist future nail trims because of this painful experience.

PROCEDURE for Expressing the Anal Sacs

Anal sacs are located on either side of the anus, just inside the lining of the rectum, approximately in the 4 o'clock and 8 o'clock positions Each sac has a duct that opens into the rectum, and the contents are normally expressed during defecation.

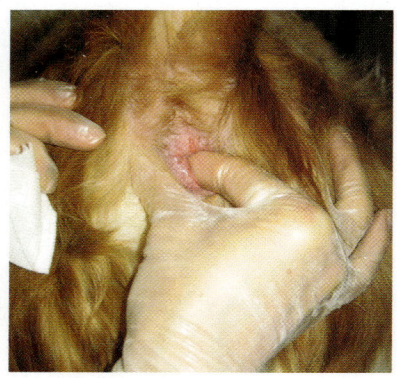

▲ **FIGURE 1-41** Manual expression of anal sacs is sometimes required for dogs.

The contents of anal sacs are foul-smelling and are thought to function primarily in territorial marking. Occasionally, dogs are unable to empty their anal sacs normally and they have to be expressed manually. When the glands are not emptied, they cause discomfort, become distended and inflamed, and occasionally rupture. Anal sacs can be emptied externally by putting gentle pressure on either side of the anus. This method does not completely empty the anal sacs and sometimes causes bruising to the delicate tissue and further blocks the ducts. Internal expression is preferred. (Figure 1-41)

Cats rarely have a problem with normal emptying of anal sacs. When problems are diagnosed, treatment usually requires general anesthesia and surgery to completely remove the anal sacs. Anal sac removal is also sometimes performed in dogs if the anal sacs are abscessed or have formed a **fistula**, an abnormal passage that can develop during impaction or from an abscess.

TECHNICAL ACTION	RATIONALE/AMPLIFICATION
1. Have an assistant restrain the dog in the standing position with the tail held erect.	
2. Put on disposable gloves.	
3. Apply a liberal amount of water-soluble lubricating jelly to the gloves, especially the index finger of the hand you will be using.	
4. Gently insert the first joint of the index finger into the rectum to locate the anal sac (refer to Figure 1-41).	
5. With light to moderate pressure, use the finger to push the contents toward the anus.	5a. Gently "milk" the contents of the anal sac between your inserted index finger and outside thumb into the anal sphincter opening. Do not squeeze.
6. Use your other hand to hold gauze pads to absorb the expressed fluid.	6a. Anal sac contents are normally brown and have a very bad smell. Using gauze or paper towels not only allows you to somewhat contain the odor but also to examine the contents for any abnormalities, such as blood or traces of mucous and pus.
7. Repeat the process with the other anal sac.	
8. Remove the gloves by turning them inside out with the soiled gauze pads inside. Tie the gloves in a knot and dispose of them correctly.	8a. This helps minimize the odor from permeating the exam or treatment room.
9. Thoroughly clean the area around the rectum with warm, soapy water, and rinse completely with clean warm water.	

PROCEDURE for Ear Cleaning

Ear cleaning may be necessary if there is a build-up of **cerumen** (ear wax) or other debris that has entered the ear canal. It is important to understand the anatomy of the ear canal to avoid damage to the tympanic membrane or other delicate ear structures (Figure 1-42).

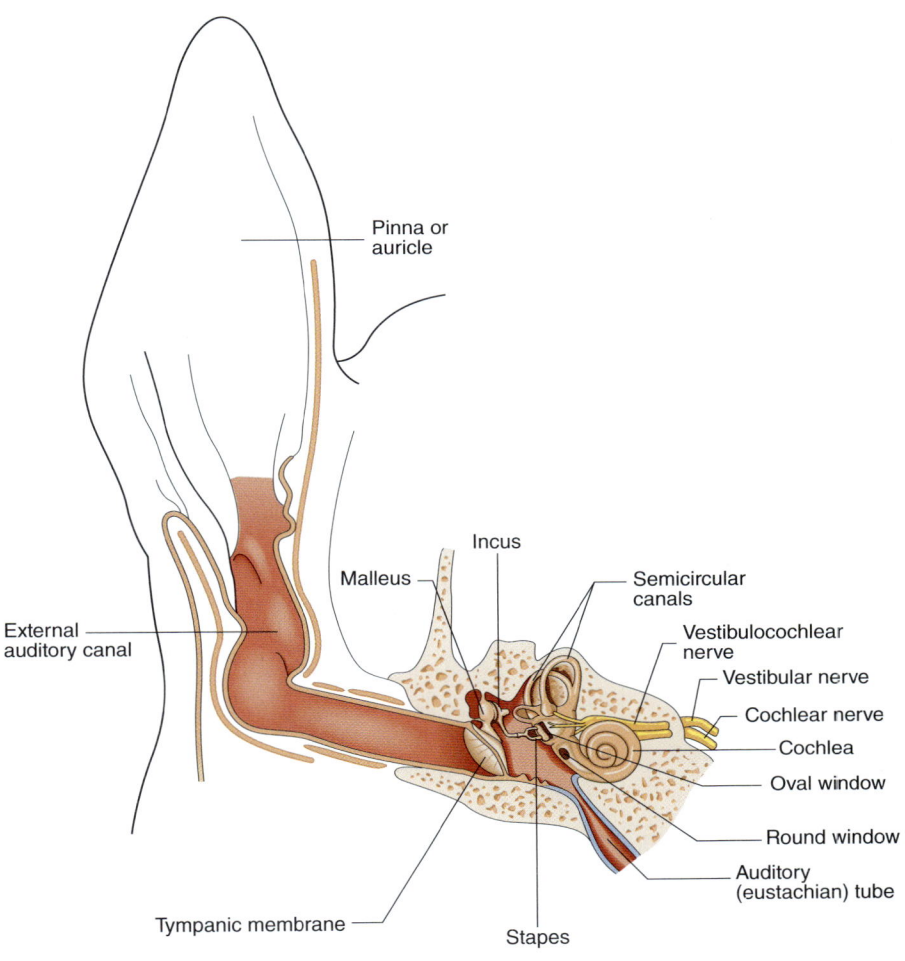

Pinna or auricle

Incus

Malleus

Semicircular canals

Vestibulocochlear nerve

Vestibular nerve

Cochlear nerve

Cochlea

Oval window

Round window

Auditory (eustachian) tube

Stapes

Tympanic membrane

External auditory canal

▲ **FIGURE 1-42** Anatomy of a dog's ear.

TECHNICAL ACTION	**RATIONALE/AMPLIFICATION**

1. Have an assistant restrain the patient in a sitting position or in sternal recumbency.

2. Examine each ear with the otoscope (**Figure 1-43**).

   2a. This is to see if the eardrum (tympanic membrane) is inflamed or ruptured. Most cleaning solutions will severely damage the inner ear if the tympanic membrane is not intact. If it appears damaged or ruptured, use only sterile saline as a cleaning agent.

   2b. Use a clean cone for the otoscope to examine each ear, to avoid contaminating a healthy ear from one that may be infected. Clean and disinfect the ear cones thoroughly after use.

   2c. If cultures are required, samples should be obtained now, prior to cleaning the ears.

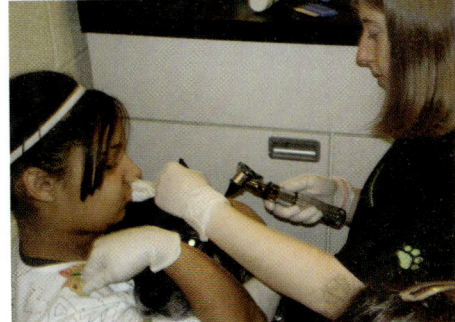

▲ **FIGURE 1-43** The restrainer must have good control of the dog's head when using an otoscope to examine a dog's inner ear.

(Continues)

TECHNICAL ACTION	RATIONALE/AMPLIFICATION
3. Note the condition and all abnormal findings. If the debris or odor is suspect, use a cotton-tipped swab and smear the material onto a microscope slide and set aside for further examination.	
4. Using the hemostats, gently remove all hair from the inside of the external ear canal.	**4a.** Only a few hairs should be pulled, or plucked, at one time. This can be a painful process, especially for a patient with an already inflamed or infected ear.
5. Elevate the nose and tip the head slightly to the side, away from you. Hold the ear flap, the **pinna**, up and pull it gently upward and outward for greater visibility (**Figure 1-44**).	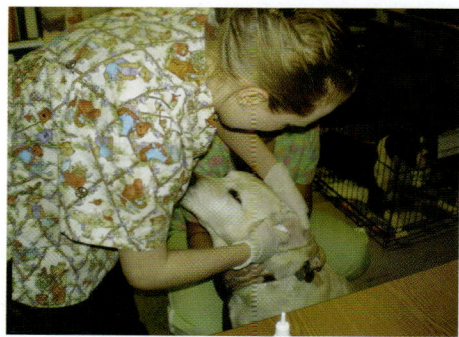 ▲ **FIGURE 1-44** Correct positioning of the dog's ear for cleaning. Most dogs do not object to routine ear cleaning, and minimal restraint is required.
6. Instill the cleaning solution into the ear with the bulb syringe (have a basin ready to collect the excess).	
7. Massage the base of the ear with your thumb and forefingers to loosen debris and wax.	
8. Wipe out excess with cotton balls.	**8a.** The dog's normal behavior is to shake its head, and it should be allowed to do this. It helps loosen deep-seated debris.
9. Repeat steps 6 through 8 until clean.	
10. Use cotton-tipped swabs to clean the folds of the pinna. Never insert them into the ear canal.	**10a.** Cotton-tipped swabs should never be inserted into the ear canal. If the patient shakes its head or bends its head into the swab, there is the danger of a swab piercing the tympanic membrane.
	10b. You may also use a finger, wrapped in cotton or gauze, to wipe out the ear. This avoids any potential of damaging the tympanic membrane.
11. Re-examine the ear visually and with the otoscope.	**11a.** Be sure to note any area of concern, and direct it to the attention of the veterinarian for further examination.

Patient History

© Liliya Kulianionak/Shutterstock.com

OBJECTIVES

Upon completion of this unit, the reader should be able to:

▶ Demonstrate effective communication skills

▶ Obtain the most complete history of the patient as possible

KEY TERMS

body language

microchip

open-ended question

paraphrase

presenting complaint

signalment

GUIDELINES FOR COMPLETING A PATIENT HISTORY

Obtaining an accurate patient history is an important aspect of total patient care. The information obtained from the client and completed accurately by the technician is the basis for all veterinary visits and treatment plans. If is often referred to during a consultation and, in some instances, can even lead to a potential diagnosis and provides the veterinarian with a solid foundation of data from which to proceed. Like all medical records, it is also a legal document and contains current and confidential client information.

PURPOSE

- To establish a base of trust and caring with the client and patient
- To listen and interpret what the client is describing
- To accurately record the information
- To provide the veterinarian and staff with essential information

COMPLICATIONS

- Inability to communicate effectively with the client
- Complete patient history unknown to client

EQUIPMENT

- Client/patient history form
- Black pen
- Clipboard
- (alternatively) Computer-generated interview form

PROCEDURE for Obtaining a Patient History

The veterinary technician must use good communication skills, including carefully listening to the client, accurately documenting information, and clearly communicating with other veterinary staff.

TECHNICAL ACTION	RATIONALE/AMPLIFICATION
1. Introduce yourself to the client and the patient.	1a. Establishing a rapport with the client and patient is important for client trust and communication.
2. Be professional, friendly, open, and caring.	2a. Try to avoid professional jargon and medical terminology; many clients do not understand, and some clients may be reluctant to ask what something means.
	2b. Ask **open-ended questions**, those that require more of an answer than a simple "yes" or "no."

(Continues)

TECHNICAL ACTION	RATIONALE/AMPLIFICATION
	2c. Do not interrupt the client. Listen carefully and **paraphrase** what the client has told you. This way, there is less confusion. Paraphrasing means repeating back what the client has said but using different words to rephrase the issue, and it sometimes allows the interviewer to sort out what is relevant and what is not.
	2d. Pay close attention to the client's **body language**. When people are stressed, they sometimes say one thing but their body language and expressions deliver a different message.
3. Explain why you need the information.	
4. Confirm owner details.	
5. Obtain or confirm patient **signalment**, the breed, age, sex, color, and reproductive status.	**5a.** Signalment refers to the patient's details and includes breed, sex (intact or neutered/spayed) age, color, distinguishing markings, and presence of an implanted **microchip**, which contains owner information.
	5b. Many rescued/shelter pets are now routinely microchipped. If the owner is unaware of a chip, use a microchip scanner to determine if a chip is present. The chips are small, about the size of a grain of rice, and are inserted under the skin in the area of the neck and shoulders.
6. Determine the reason for the visit, the **presenting complaint**.	**6a.** Allow the client to describe the concern. He or she may have an opinion or a "diagnosis," but never attempt to correct this presumed diagnosis or interject your own conclusions. The veterinarian will address these issues.
7. Complete all sections on the history form. (Refer to **Figure 1-45**, sample patient history form)	**7a.** Remember that this is also a legal document. Leaving an area blank could mean that something was "skipped over" and potentially could lead to a charge of professional neglect. If the question does not apply to a patient, enter "n/a" (not applicable) to show that it has been addressed but is not relevant to this specific patient. If using a hard copy and pen, do not scribble over or white-out mistakes. Instead, draw a line through the incorrect information, initial it, and enter the correct information immediately afterward.

DATE _____ CASE NUMBER _____

CLIENT/PATIENT INFORMATION FORM

Please provide the following information for our records: **PLEASE PRINT!**

OWNER INFORMATION

Owner's Name

Street Address

City/State	Zip Code	Country
Telephone (Include Area Code)	Home	Business
Drivers's License Number	Place of Employment	How Long Employed?

ANIMAL INFORMATION

Animal Species (Dog, Cat, Other)		Breed
Animal's Name	Sex	Has the animal been altered? ☐ Yes ☐ No

Color	Birth Date

REFERRAL INFORMATION

Where you referred by a veterinarian?

	☐ YES	☐ NO	If so, complete the following information.

Veterinarian's Name	Phone

Street Address

City/State	Zip Code

You will be advised of estimated cost and anticipated procedures. Please feel free to discuss the proposed treatment and any costs with the veterinarian. A minimum deposit of 50% of the initial estimated charges will be required for hospitalization of the patient.

STATEMENT OF OWNERSHIP AND CONSENT: I am the owner of the above-described animal, or have authorization from the owner to consent to its treatment.

   I hereby authorize the performance of professionally accepted diagnostic, therapeutic, anesthetic, and surgical procedures necessary for its treatment.
   I accept financial responsibility for these services.
   I have read the above consent and understand why these procedures may be necessary. I have also been told of the possible complications and alternatives to the anticipated procedures.

PAYMENT CHOICE: ☐ Cash   ☐ Check   ☐ Bank Card

SIGNATURE (Owner/Agent)	DATE

▲ **FIGURE 1-45** Sample client/patient Information Form.

Physical Examination

© Lilliya Kulianionak/Shutterstock.com

OBJECTIVES

Upon completion of this unit, the reader should be able to:

▶ Complete a physical examination using the system-by-system approach

▶ Demonstrate use of the equipment

▶ Perform a complete physical examination

▶ Accurately record and understand the findings

▶ Understand the importance of the physical examination

KEY TERMS

alopecia	diastolic	popliteal
arrhythmia	dypsnea	prognathism
atrophy	gingivitis	pulse deficit
auscultate	hematoma	stridor
axillary	inguinal	submandibular
bite	integument	systolic
bracheocephalic	intercostal	thorax
capillary refill time (CRT)	musculoskeletal	thrombosis
	overbite	underbite
deciduous teeth	palpate	urogenital

GUIDELINES FOR THE PHYSICAL EXAMINATION

The physical examination follows completion of the patient history. Together, they present a profile of the patient on that specific day. Prior to the physical examination, review the history not only for physical concerns and past medical problems but also for behavioral concerns. The physical examination form should be followed closely, as it provides a system-by-system evaluation. All findings should be accurately recorded, and no areas should be left blank. Blank spaces could indicate that something may have been missed. If an area of the examination form does not apply to the patient, always indicate this by recording "n/a" (not applicable).

The same procedure that is used as with errors on a patient history applies to the physical examination form as well. Errors in handwritten notes should not be scratched out or scribbled over. Instead, draw a line through the incorrect information, initial it, and re-enter the correct information. This indicates that a mistake was made, recognized, and corrected, rather than an entry that could be interpreted as being disguised or covered up.

The veterinary technician should always remain professional and friendly while conducting the examination and not hesitate to ask for an additional restrainer if required. It is important to keep both the patient and the owner relaxed and comfortable with the procedure, and asking for assistance with a reluctant or fractious patient is far easier than trying to struggle with the animal on your own.

Although it is normal for a client to ask questions during the examination, you should never express an opinion on the findings, either by your facial expression and body language or by vocalizing a concern you may have. Reassure the worried client that the veterinarian will discuss all concerns there may be when he or she reviews the findings. Under *no* circumstances should the technician even suggest a diagnosis even though a given finding may be obvious or indicative of a disease or area of concern.

PURPOSE

- To evaluate and record the patient's overall health
- To examine each body system to determine and record normal and abnormal findings
- To establish a baseline of findings for future reference
- To provide the veterinarian with the most complete information during the physical examination to assist in a diagnosis and/or recommendations for further evaluation

COMPLICATIONS

- An aggressive or frightened patient
- Poor restraint technique
- Examiner unfamiliar or unsure how to use the equipment
- Poor rapport with client and patient
- Failure to have all equipment and anticipated equipment organized and at hand

EQUIPMENT

- Completed patient history and the examination form
- Slip leash and other restraint equipment (muzzles, towels)

- Quiet area/examination room to conduct the examination
- Assistant for restraint
- Method of recording findings, either computer-generated form or hard copy (black pen)
- Disposable examination gloves
- Paper towels
- Thermometer
- Sterile lubricant
- Stethoscope
- Otoscope
- Watch or clock with secondhand
- Penlight
- Scale
- Alcohol
- Gauze pads

PROCEDURE for a Physical Examination

The veterinary technician's approach for the physical examination should be consistent every time it is performed. A consistent approach will ensure that nothing is missed.

TECHNICAL ACTION	RATIONALE/AMPLIFICATION
1. Begin the assessment from the moment you enter the room, and establish a calm and friendly relationship with the client and the patient.	1a. 2a. 3a. The initial examination should begin at the moment you see the patient. Many components of the examination rely on your ability to assess the patient and evaluate areas that may be of concern. This includes the animal's behavior, its awareness of the surroundings, and potential for aggression. Review the file and patient history, and note any remarks or warnings that the animal may bite. Remaining calm and friendly relaxes the patient and the owner. Companion animals are highly aware of subtle signs of anxiety from the owner's body language and tone of voice. It is just as important to keep your own body language nonthreatening and calming.
2. Observe the patient's demeanor and attitude, and note your general impressions.	2b. Walking the patient around may not be necessary if you observed the patient when the owner walked it into the room. If you note areas of concern, limping, awkwardness of gait, a head tilt, or anything abnormal, these concerns should be investigated further by reviewing the history on the medical chart, in discussion with the client, and by your own examination.
3. Prior to any hands-on examination, allow the patient time to become familiar with the surroundings and with you.	

(Continues)

TECHNICAL ACTION	RATIONALE/AMPLIFICATION

4. Obtain the weight of the patient both in pounds/ounces and kilograms.

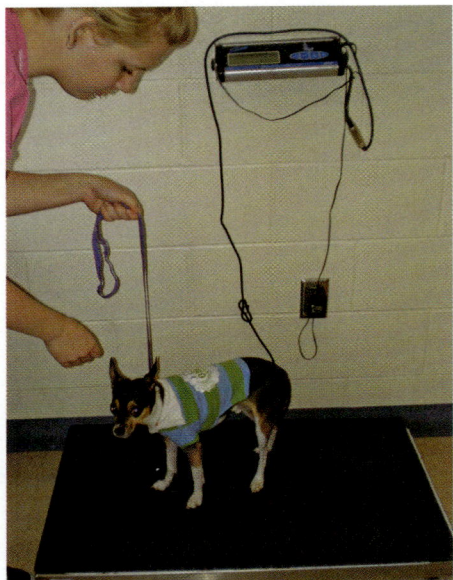

▲ **FIGURE 1-46** Dogs are weighed with a platform scale.

4a. While this is often done during check-in, always confirm that the weight has been recorded. Dogs are weighed with a platform scale (**Figure 1-46**), and cats usually are weighed with an infant scale (**Figure 1-47**).

4b. Medications are calculated according to body weight— for example, milligrams of drug per kilograms of body weight—so it is important to record the weight in kilograms. (A kilogram equals 2.2 pounds). Clients usually ask how much their pets weigh, and it often is easier for them to understand pound weight, so you can tell them clearly with both measurements. For example: "Quiggy weighs slightly over 29 kilograms, or approximately 64 pounds."

4c. To obtain the weight in kilograms, divide the pound weight by 2.2. Scales often give the weight in both pounds and kilograms according to the setting chosen on the scale.

▲ **FIGURE 1-47** Cats are weighed with an infant scale.

5. Begin with the head: eyes, ears, and nose.

5a. Check the head for symmetry, and palpate for any unusual lumps or bumps. Both eyes should be clear, bright, and free of discharge. Cloudiness is often caused by cataracts and may be present in one or both eyes. Check to see that both pupils are the same size, and with the penlight test for a normal pupil response. With the light, the pupil should contract and then dilate when the light is removed.

(Continues)

TECHNICAL ACTION	RATIONALE/AMPLIFICATION

5b. Examine the pinna of the ear (the ear flap) both inside and out. Check for areas of redness or scratching, excessive matted hair with a waxy buildup and odor. (Some breeds, Poodles, for example, normally have a lot of hair within the ear.) Hold both ears in your hands to see if they feel hot, or one hotter than the other. Palpate the pinna to check for a **hematoma** and other "bumps and lumps." Hematomas are clots of blood that form when the ear has been bruised. They usually are at the tip of pinna but can form anywhere within the pinna. Head shaking can indicate either an infection or a foreign body within the ear canal. If the patient holds its head to one side, that also indicates a problem. Usually the head will be held to the same side as the affected ear.

With the otoscope, check the inner ear, being careful not to advance the otoscope too deeply. It wise to have an assistant restrain the patient's head to prevent the otoscope from inadvertently penetrating too deeply and damaging the tympanic membrane.

6. Examine the oral cavity: mouth, teeth, tongue, and mucous membranes.

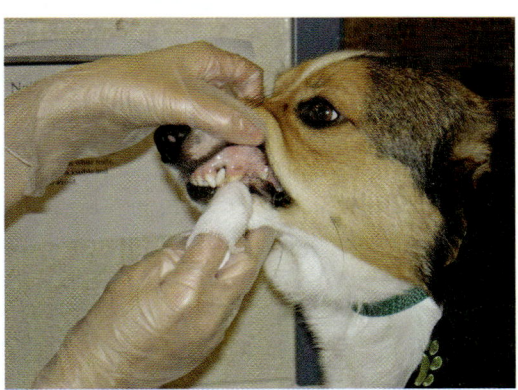

▲ **FIGURE 1-48** The patient's CRT and mucous membrane color can be checked by lifting the upper lip to observe mucous membrane color and during same time perform a CRT test.

6a. If the patient has shown no signs of aggression and you think it is safe do to so, examine the mouth by first placing one hand over the head and using the other to lift the lips (**Figure 1-48**). Check the teeth, and note the amount of tarter present, signs of **gingivitis** (inflamed gums), and any broken or missing teeth. Check for **deciduous teeth**, the baby teeth that have been retained. Note any peculiar odor coming from the patient's mouth.

6b. With the mouth closed, evaluate the **bite**—how the top and bottom incisors meet. There may be an **overbite**, where the upper teeth extend beyond the lower teeth, or an **underbite,** where the lower teeth extend beyond the upper teeth. This is more common in dogs than cats. Some breeds, Bulldogs, for example, normally exhibit **prognathism** (protrusion of the lower jaw), but in most breeds the upper and lower incisors should be aligned.

6c. Check the mucous membranes for color and **capillary refill time (CRT)**. Mucous membrane color should be determined from an unpigmented area of the gums. (Refer to the description of mucous membrane colors other than the normal, healthy pink.)

6d. The CRT is assessed by pressing your fingertip on an area of the gums. It will blanch, or turn white. CRT reflects perfusion, or how quickly the blood returns to the tiny capillaries. Normally, CRT is under 2 seconds. The mucous membranes should feel moist, not dry or tacky, and this observation can help evaluate the patient's hydration status. (Refer to examination of the integument for further amplification of hydration status.)

(Continues)

MUCOUS MEMBRANE COLORS

Blue: very low oxygen levels; emergency!
Pale/white: very low blood platelets; emergency!
Icteric (yellow): can indicate liver disease
Brick-red: toxic or septic shock; emergency!

DARK PIGMENTATION OF MUCOUS MEMBRANES

It is not unusual for dogs to have areas of black pigment on the tongue and gums. This does not indicate disease, a breed type, or a positive determination of a dog with mixed breeding. The exception to this is the Chow Chow. This breed not only has blue-black pigmented gums, but also a black tongue, and it is characteristic of the breed.

TECHNICAL ACTION	RATIONALE/AMPLIFICATION
7. Evaluate the **thorax** (the chest). Count the respiration rate and observe the breathing pattern.	7a. Observe the thorax, and count the number of breaths. A complete breath is one inspiration (inhale) and one expiration (exhale). The simplest way to obtain the respiration rate is to count for 15 seconds and multiply that number by 4, which gives you the respiraton rate (RR) for one minute. It is recorded as RR. Normal respiration rate in an adult dog is 12–20 breaths per minute. Normal respiration rate for cats is slightly higher, 12–30 breaths per minute. Respiration rate can be influenced by pain, disease, trauma, excitability, and stress. If the patient is panting, record it as such and wait until the patient has settled down and you can make a more accurate observation. Cats often pant when stressed, especially during a car ride if they are not accustomed to travel. Let them settle quietly, and gently reassure them with your voice and touch.
	7b. Open-mouth breathing is different from panting. A patient with open mouth and labored breathing, termed **dyspnea**, is in severe respiratory distress and the veterinarian should be notified immediately.
	7c. Audible respiratory sounds, **stridor** (loud breathing), and rapid, shallow breaths should be brought immediately to the attention of the veterinarian. Using the stethoscope, record any wheezing or crackles you hear. (Crackles sound like crumpled cellophane.)

(Continues)

TECHNICAL ACTION	RATIONALE/AMPLIFICATION

7d. Consider all other signs, including breed. Pugs, Pekingese, Boxers, Bulldogs, and Persian cats, These are representative of **brachiocephalic** breeds with foreshortened faces and they frequently have genetic abnormalities of the upper respiratory tract and breathing sounds are audible. Use the stethoscope to determine where the breath sounds originate. Breath sounds can indicate not only an emergency but also disease within the respiratory system or respiratory compromise from other organ systems, such as the heart.

8. Using a stethoscope, obtain the heart rate, and at the same time, place two fingers on the femoral artery to obtain the pulse rate. **Auscultate** (listen with the stethoscope) for any abnormal heart sounds.

8a. To obtain the heart rate (HR), place the stethoscope on the left side of the thorax. The heart beat can be located between the fourth and sixth **intercostal** space (between the fourth and sixth ribs). Normal heart rates for the dog are 70–160 beats per minute (bpm). Normal heart rates for the cat are 150–210 beats per minute. The count can be taken by listening to the heartbeat for 15 seconds and multiplying by 4. After attaining the heart rate, continue to listen for an **arrhythmia,** or irregular heartbeat, and any abnormal sounds or muffled heart beats. Always move the stethoscope to at least three different areas to determine where the heart sounds are heard most clearly.

8b. The pulse rate can be taken at the same time as the heart rate. A pulse occurs between the heart contracting, called **systolic,** and the heart refilling, **diastolic**. The pulse can be easily palpated by placing two fingers over any artery; however, the most commonly used site is the femoral artery, located on the inner thigh and felt best near the groin.

8c. Pulse rate and strength are assessed simultaneously with the heart rate to determine if there is a **pulse deficit**, when the pulse rate differs from the heart rate. In cats, the pulse should be checked in both femoral arteries simultaneously. With the cat standing, use the first and second fingers of both hands to locate the pulse and determine if both are strong and synchronized. A pulse deficit in either femoral artery may be indicative of a **thrombosis** or clot, especially in a cat exhibiting a hind limb abnormality not caused by trauma.

9. Auscultate the trachea (upper airway) for abnormal breath sounds.

9a. Listen first to the airway for audible sounds without using the stethoscope. Listen to the air moving in and out of the nares and the mouth. Then use the stethoscope to help determine where the sounds originate.

(Continues)

TECHNICAL ACTION	RATIONALE/AMPLIFICATION
10. Examine the entire **integument** system: hair coat, skin, foot pads, and nails.	**10a.** Examine and evaluate the condition of the skin and hair coat. Look for areas of hair loss, **alopecia**, dry and flaky skin, and areas of redness and sores. Heavily matted coats frequently have open sores under the mats, or there may be "hot spots" where the skin has deteriorated. Check for the presence of external parasites: fleas, ticks, and mites. Small round-appearing areas of alopecia may be caused by ringworm, a skin fungus that can be transferred to humans, one reason that gloves should always be worn during the physical examination.
	10b. Hydration status also can be evaluated by tenting the skin over the shoulders. The skin should not remain tented but instead should return immediately when released. This may be recorded as 5%, meaning that dehydration is not evident. (Refer to Fluid Therapy for further evaluation and recording of hydration status.)
	10c. Carefully check the nails and nail beds, and note the presence and condition of dewclaws. Examine the foot pads for signs of dryness and cracking, open sores or areas of excessive wear. Check the skin between each toe on each foot, and look for growths, foreign objects, cuts, and redness. The area between the toes often accumulates debris and mats of hair, especially in long-haired dogs. Look for evidence that would indicate that the patient has been chewing at the feet and foot pads and, if possible, determine the source of irritation.
11. During the examination of the integument, **palpate** the peripheral lymph nodes. Easily palpated lymph nodes are **submandibular** (beneath the jaw), **axillary** (armpit area), **inguinal** (groin), and **popliteal** (medial area of rear limbs).	**11a.** Swollen lymph glands in any or all palpated locations are abnormal and may indicate any one of several disease processes or infection.
12. Evaluate the **musculoskeletal** system: muscles, bones, and joints, for any signs of lameness, swelling or pain and **atrophy** (muscle wastage).	**12a.** Evaluate muscle tone and look for signs of atrophy (muscle that has lost tone).
	12b. Flex the joints of each leg. Do not attempt to force movement; the patient may be arthritic or have other joint problems. Forcing movement not only causes pain but could further contribute to the problem.
	12c. Palpate each leg from shoulder to the toes to determine if there are any hard lumps, areas of swelling, heat, or pain.
13. Palpate areas of the abdomen, gently feeling with your hands for any signs of tenderness, abnormal swellings (hard or soft) in the stomach and intestines. With experience, other organs also may be palpated in the same manner.	

(Continues)

TECHNICAL ACTION	RATIONALE/AMPLIFICATION
14. Examine the **urogenital** system: bladder, external genitalia, mammary glands, and rectum.	**14a.** A distended bladder can be easily palpated. It may be that the patient has not voided recently, is too nervous to do so, or that there may be a blockage. Make a note of when the client last observed the dog urinating or when the cat last used the litter box. If the client reports that the male cat is not using the litter box, strains, or vocalizes, this could be an indication of a "blocked tom." Male cats often present problems with urination. Refer to the patient history and the chief complaint. Do not put any pressure on an extended bladder.
	14b. The external genitalia should be clean and free from odors and evidence of discharge with no signs of licking or chewing. If the patient is an intact bitch (female dog), determine her estrus cycle or when the client last observed that she was "in heat" or if she ever had a litter of pups. This may be the first estrus, so refer to signalment for age and reproductive status. Examine the sheath of a male dog for any sign of discharge, irritation, and evidence of licking and chewing. If un-neutered, palpate both testicles. It is important to note if only one testicle is present.
	14c. Palpate the mammary glands for areas of hardness, heat, and difference in size. This applies to both males and females.
	14d. The outer rectal area should be clean and free of fecal material with no abnormal swelling, protrusion, or obvious signs of irritation and itching.

UNIT SIX

Care of the Hospitalized Patient

OBJECTIVES

Upon completion of this unit, the reader should be able to:

▶ Identify the importance of keeping the cages and kennel areas clean

▶ Observe and record abnormal findings

▶ Correctly use a whiteboard for patient treatments

▶ Understand commonly used abbreviations for treatment protocol

KEY TERMS

decubitus ulcer

iatrogenic

Toxoplasmosis

treatment board

urine scald

CAGE AND KENNEL MAINTENANCE

Good nursing care begins by providing the patient with a clean, dry, and comfortable area. Patients should be checked frequently for urine and feces soiling, to ensure that the bedding is dry, that there is fresh water available, and that uneaten food has been removed and recorded. This routine maintenance also provides nursing staff the opportunity to evaluate the patient's attitude and patency of sutures, fluid lines, and bandages.

Soiled cages should be cleaned immediately: the ceiling, sides, floor, and bars washed down and fresh bedding and water supplied. Many **iatrogenic** conditions—those that occur as a result of poor nursing skills—can be avoided with proper care. For example, one problem that can easily develop and complicate recovery is **urine scald**; when the patient is left in a urine-soaked cage, the skin and underbelly become inflamed. Another problem that can occur, especially in a recumbent patient, is a **decubitus ulcer**, or pressure sore. Pressure sores can be avoided by providing ample, soft bedding and frequently turning a recumbent patient. Patients should be turned every 2 hours and always placed facing the cage door. The patient not only is monitored more easily, but this provides the patient with more stimulus and comfort than having to look at the back of the cage. A soft rolled towel can be placed between the two hind legs for greater comfort to the patient, and towels should be replaced promptly if they become damp or soiled. Remember: Total patient care is more than just the administration of medications and providing food and water.

Many times, as a result of frequent opening, a latch may become loose and the door doesn't close properly. Be sure to check each cage or kennel latch to make sure that it is secure. You certainly would not want a patient to roll against the front of the cage and have the door pop open. Serious injury, even death, could occur if a patient falls to the floor and fractures a limb or suffers a broken neck.

All patients, admitted for any reason, should be discharged in better condition than when they were admitted. Clean and fresh patients reflect well on the entire hospital, but especially on those with whom their care has been entrusted.

GUIDELINES FOR CAGE AND KENNEL CLEANING AND MAINTENANCE

All equipment and supplies needed should be gathered together and ready before beginning. It often helps to have a cart prepared and used only for cleaning supplies so it is ready when required. Always begin with the healthiest of patients to avoid transferring pathogens as much as possible. Patients with a condition that may be contagious should be left for last. Patients in isolation units are cleaned last, often with the use of equipment dedicated for use only in the isolation unit, including gloves, gowns, caps, shoe covers, and masks. Note that some patients in the isolation ward are likely to require more frequent cage care, as often they are very ill and are not taken out for normal elimination.

Patients that are soiled should be spot-cleaned and dried before returning them to a clean cage. A clean cage should be ready to receive a patient while the assigned kennel or cage is being cleaned. The holding cage should be cleaned and wiped down after every occupant. Gloves should be worn and disposed of with the other material from the cage. Hands should always be washed and new gloves worn between handling patients. Soiled scrub tops or aprons should be changed immediately.

PURPOSE

- To remove all traces of blood, feces, and urine from the cage/kennel and doors
- To replace bedding, newspapers, and blankets
- To provide clean food and water bowls
- To keep the patient in an environment that is clean and healthy
- To promote the patient's health and wellness
- To observe and record the patient's status

COMPLICATIONS

- A severely injured patient
- A critically ill patient
- A patient recovering from surgery
- A large or aggressive patient

EQUIPMENT

- Box of disposable gloves
- Rolls of paper towels
- Bucket with warm soapy water
- Scrubbing brush that is easily disinfected after each use
- Spray bottle of disinfectant (for safe and approved use in cage cleaning)
- Trash bags
- Newspapers
- Clean blankets and towels
- Phillips head screwdriver
- Flat-bladed scraper

PROCEDURE for Cleaning a Dog Cage

TECHNICAL ACTION	RATIONALE/AMPLIFICATION
1. Put on a pair of disposable gloves.	
2. If cage is occupied, move the patient to a ready-prepared holding cage.	
3. Remove food and water dishes and let soak in hot, soapy water in the sink.	
4. Remove the soft bedding, towels, and blankets, and place in a laundry bag.	
5. Fold the soiled newspaper layers inward (toward the center) so they form a package that can be easily placed in the disposal bag.	
6. Apply a liberal amount of hot, soapy water and allow it to loosen adhered fecal matter or matted down paper.	

(Continues)

TECHNICAL ACTION	RATIONALE/AMPLIFICATION

7. If necessary, use the scraper to lift the material and deposit it directly into the disposal bag.

8. Begin by cleaning the top of the cage (ceiling), the back and both sides, rinsing frequently.

9. Clean the bars of the door making sure that there is no debris between the bars, latches, and hinge. Cage doors may be removed for easier cleaning.

10. Wipe all cage surfaces with soapy water and sponge out with clean water (**Figure 1-49**).

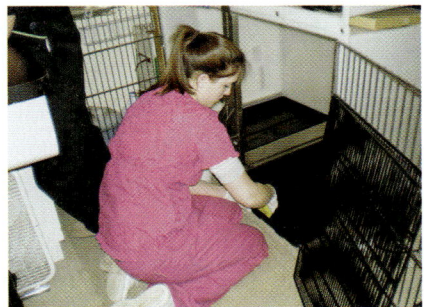

▲ **FIGURE 1-49** Patient cages and kennels should be routinely cleaned and disinfected. This is an important aspect of total nursing care.

11. Spray all areas of the cage with the disinfectant solution and allow it to stand, according to the manufacturer's directions.

11a. Different disinfectant solutions require various contact time for maximum effect. Always follow the manufacturer's direction with regard to contact time and dilution recommendations.

12. Wipe the surface dry with clean, disposable paper towels.

13. Put down a layer of newspaper and other soft bedding as needed.

14. Replace the water dish with fresh, clean water.

15. Check that all screws, door hinges, or gates are secure and tighten with the screw driver if necessary to ensure a secure latch.

15a. Occasionally, a large dog can be placed temporarily in an empty ward or a consultation room, but *only* when there is no other holding space available. Sometimes the dog may become frantic at this sudden isolation. Always monitor the dog closely and check for damage to the room and door handle (**Figure 1-50**).

◀ **FIGURE 1-50** The handle of this door was chewed off by a frantic dog in an attempt to get out of the enclosed room in a veterinary hospital. Dogs should not be left alone and unsupervised while waiting for a kennel to become available.

16. Return the patient to the assigned cage.

17. Remove the gloves and dispose of correctly.

(Continues)

TECHNICAL ACTION	RATIONALE/AMPLIFICATION
18. Wipe down the holding cage with disinfectant or thoroughly clean if necessary.	
19. Record any unusual findings in the patient record, and advise the veterinarian if there is a change or any area of concern.	19a. Feces should be observed and recorded. Describe the color, consistency, volume, and any other abnormality observed. Some examples are: yellow, green, coffee-ground appearance, black and tarry, presence of fresh blood, mucoid, watery, projectile, foul-smelling.

PROCEDURE for Cleaning a Cat Cage

Follow the procedure for cleaning a dog cage, above. In addition, cat cages should always contain a litter box, and it is important to keep them cleaned throughout the day, regardless of an established routine. Many feline patients prefer to curl up in the litter box in preference to other bedding provided, as it seems to make them feel less exposed and more secure. Also, it is an important step in preventing undesirable odors and diseases. When cleaning the litter box, always wear gloves and wash your hands afterward (**Figure 1-51**).

Pregnant women should not clean litter boxes because of the potential of **toxoplasmosis**, *a zoonotic microscopic parasite. Exposure to Toxoplasma gondii can cause birth defects.*

Photograph by Kathy Nuttal, courtesy of Riverton Veterinary Hospital

◀ **FIGURE 1-51** When cleaning cat litter boxes gloves should always be worn.

GUIDELINES FOR PERFORMING TREATMENTS OF THE HOSPITALIZED PATIENT

Prior to performing any treatments, the **treatment board** should be consulted. The treatment board is not unlike a flowchart and when used correctly, it ensures that each patient receives scheduled treatments and all concerns for each patient are addressed during the day. This is usually a large whiteboard in the treatment area so all attending staff members can see at a glance the patients that are admitted and the procedures and treatments to be done. Some practices may also use a separate whiteboard for surgical patients, and these patients are usually listed in the order that the surgeries will be performed. **Figure 1-52** demonstrates the use of a whiteboard. When a procedure is completed, each square should be crossed through and initialed. In this way, nothing is left out and medications, for example, are not doubled-up or missed completely. By initializing the box, the person who administered the treatment is easily identified if there is any question.

▲ **FIGURE 1-52** A whiteboard is used to track treatments and procedures.

Collect the patient files required and compare them to the entries on the whiteboard. When a treatment has been completed, mark it off on the whiteboard, initial it, and then record the treatment in the patient file. The computer file for each patient then can be updated from these "traveling files." Remember—if you have also been attending to patient care by cage cleaning, feeding, and watering, these notes and comments should also be recorded in patient files.

UNIT SEVEN

Blood Collection

OBJECTIVES

Upon completion of this unit, the reader should be able to:

▶ Identify the anatomical location of various veins

▶ Demonstrate restraint techniques used to facilitate venipuncture

▶ Demonstrate how to perform a blood draw

▶ Identify different types of collection tubes and the purpose of each

▶ Demonstrate correct method of handling a blood sample

KEY TERMS

anticoagulant

jugular vein

occlude (occluding)

venipuncture

GUIDELINES FOR OBTAINING A BLOOD SAMPLE

Two of the most frequently required and applied skills for the veterinary technician are the abilities to restrain a patient for a blood draw and to perform the blood sample collection. This requires teamwork, communication, knowledge, and practice if it is to be performed successfully and with the minimum amount of stress to the patient. The restrainer and the person performing the blood draw must know the purpose for the blood draw, the volume of blood required, the location of the vein of choice, and which method of restraint will be used. In addition, all supplies must be ready, at-hand, prior to beginning the procedure. The team also needs to know how the blood sample is handled, which collection tubes are used, and the appropriate-sized needle and syringe. It is important to collect blood prior to administering fluids or other treatments that could affect the results. Stress to the patient during the procedure can also alter blood values.

The sites used for **venipuncture** in dogs are the cephalic vein, the *lateral* saphenous vein, and for larger amounts, the **jugular vein**. For easier access and vein visualization, the *medial* saphenous vein is used for cats instead of the lateral saphenous vein used in dogs. With the exception of a jugular venipuncture, an assistant not only restrains the patient but also is responsible for **occluding** ("holding off" the vein) and releasing the vein *prior to* syringe withdrawal and to maintain pressure at the puncture site until all bleeding stops. Because of the location of the jugular vein and the restraint technique required, the person performing the blood draw occludes the vein and releases it. The restrainer can then apply direct pressure to the puncture site. Successful blood collection requires observation and communication.

Refer to **Figures 1-53 to 1-55** for the restraint techniques required for different venipuncture sites in the dog. **Figures 1-56 to 1-58** demonstrate restraint techniques used in cats. If may be beneficial to review the unit on restraint and master these techniques with a cooperative patient without attempting a blood draw.

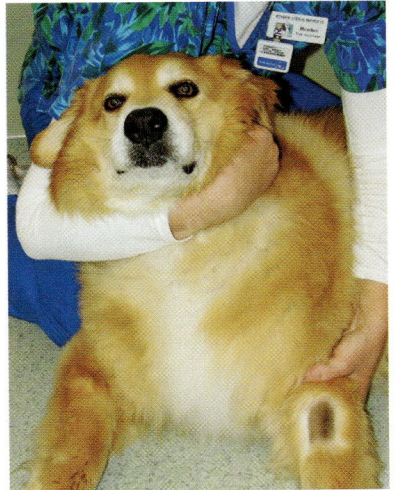

▲ **FIGURE 1-53** Restraint and positioning for a blood draw from the cephalic vein in a dog.

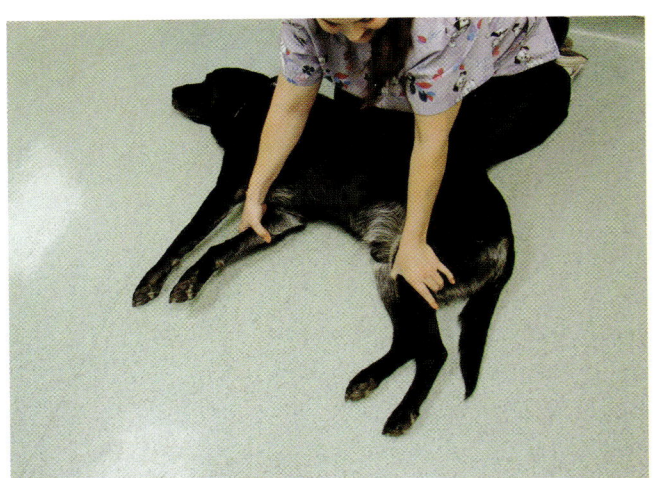

▲ **FIGURE 1-54** Restraint and positioning for a blood draw from the lateral saphenous vein in a dog.

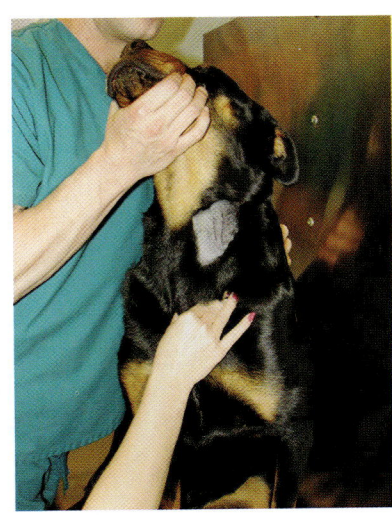

▲ **FIGURE 1-55** Restraint and positioning for a blood draw from the jugular vein in a dog.

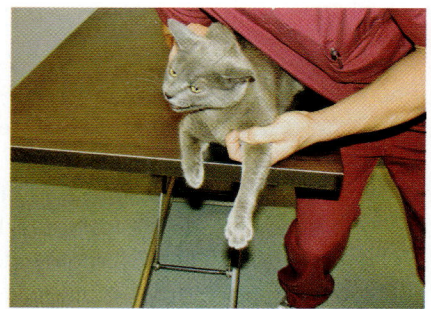

▲ **FIGURE 1-56** Restraint and positioning for a blood draw from the cephalic vein in a cat.

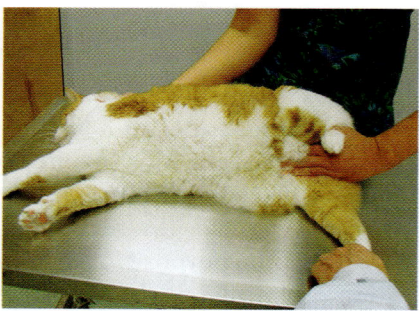

▲ **FIGURE 1-57** Restraint and positioning for a blood draw from the medial saphenous vein in a cat.

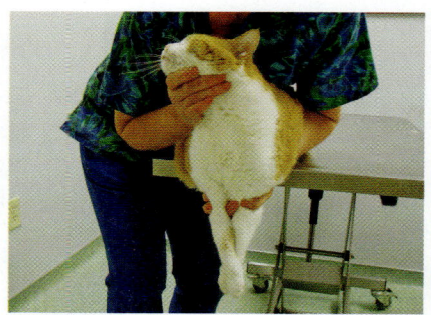

▲ **FIGURE 1-58** Restraint and positioning for a blood draw from the jugular vein in a cat.

PREPARING THE SITE FOR A BLOOD DRAW

Note that sometimes the venipuncture site is shaved to more easily visualize the vein. This is not always done, nor is it required; however, when it is, use a # 40 blade and keep the shaved area as small as possible. The edges of the shaved area should be neat and clean. The shaved patch should be wiped with alcohol to clean the site and to remove small particles of hair and dander. Alcohol also causes the vein to become more prominent and easier to visualize. If using alcohol, allow the alcohol to dry prior to the puncture. Other antimicrobial solutions (never scrubs) may be used, such as dilute betadine or dilute chlorhexadine.

PURPOSE
- To obtain a blood sample adequate for various laboratory tests

COMPLICATIONS
- Inadequate restraint
- Injury to the limbs or neck
- Hematoma at puncture site
- Inadequate sample volume
- Stress to patient (stress can alter some blood values)
- Bite injuries to the team performing the procedure

EQUIPMENT
- 20–25 gauge needle
- 3 ml syringe
- Blood collection tubes
- Indelible pen for labeling tubes
- Electric clipper with # 40 blade
- Alcohol wipes
- Cotton balls
- Strip of one-inch-wide white tape
- Patient record

COLLECTION TUBES

The type of blood analysis will determine which blood collection tube(s) should be used. Different colored tops identify the type of **anticoagulant** they contain that prevents the blood from clotting (**Figure 1-59**). Tubes should always be filled one-half to three-quarters full of blood to achieve the correct mixture of blood with the anticoagulant if there is one, and to provide an adequate volume for testing. Not all tubes contain an anticoagulant, so be sure to use the correct collection tube(s) for the tests required.

- **Purple: EDTA** (ethylenediaminetetra-acetic acid) is the most frequently used collection tube for whole blood; **CBC** (complete blood count)
- **Red:** contains no anticoagulant and is used for blood chemistries
- **Red and Black striped:** no anticoagulant and contains a visible separator gel, used for chemistries (also called a "tiger top")
- **Green:** contains **heparin**, an effective anticoagulant if sample is analyzed promptly, i.e., less than a few hours. Blood smears (blood films) should be made immediately to avoid cell clumping.
- **Turquoise** (blue) contains sodium citrate and is used mainly for coagulation studies. It should not be used for routine blood analysis (CBC).

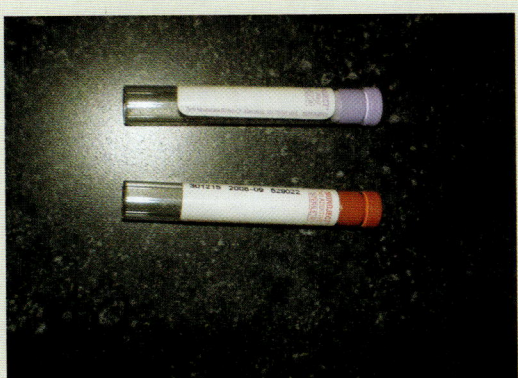

▲ **FIGURE 1-59** Two of the most frequently used blood collection tubes are a red top and a purple top. The red top tube does not have an anticoagulant, while the purple top tube contains the anticoagulant EDTA.

PROCEDURE for Obtaining a Blood Sample

TECHNICAL ACTION	RATIONALE/AMPLIFICATION
1. Confirm that it is the correct patient.	
2. Confirm tests ordered and samples required.	
3. Gather and prepare all the necessary supplies.	3a. Usually, a 3–5 ml syringe is used. Using a larger syringe can cause a vein to collapse because of the vacuum suction.

(Continues)

TECHNICAL ACTION	RATIONALE/AMPLIFICATION
	3b. Always use a new (sterile) syringe. Check the plunger by moving it back and forth a few times, to ensure that it moves freely. (Do not remove it entirely from the syringe while doing this.)
	3c. Select the appropriate-sized needle. Needle size depends upon the size of the patient and the patency of the vein. Needles of 25 to 22 gauge may be appropriate. Using a smaller needle may damage blood cells as they are pulled through a smaller lumen. A larger gauge needle may cause trauma to fragile veins. Always consider the size and age of a patient and choose a needle gauge that will fit comfortably within the vein. With cap still in place, attach it firmly to the syringe. Remove the cap and inspect the needle tip for any flaws or burs at the tip. Discard and change if necessary. Re-cap the needle to keep it sterile.
	3d. Pull a length of white adhesive tape approximately 8 inches long. Fold one end of it over so that it forms a tab. Stick the other end on the table close to the cotton balls so that the restrainer has easy access. The restrainer can use a cotton ball to apply pressure at the puncture site and wrap it with the tape. The tab should be at the end of the wrap and it provides for easy removal.
4. Determine vein of choice.	**4a.** Access to a specific vein may be limited because of an existing injury or previous blood draws that have compromised the patency of the vein. Vein choice also can be influenced by the volume of the sample required. For large amounts, the jugular would be the vein of choice. In obese patients, all veins can be difficult to palpate.
5. If necessary, shave the puncture site and clean with alcohol.	**5a.** Shaving the site provides for greater visualization, especially in an obese patient. Alcohol cleans the puncture site of hair clippings and debris.
6. Have restrainer position patient for the blood draw and occlude vein.	
7. Palpate and stabilize the vein with the opposite thumb.	
8. Introduce the needle, bevel edge up at a 25° angle from the most distal point of the occluded vein.	**8a.** The skin and the vein are punctured simultaneously. The bevel edge of the needle should be facing up. In this way, it serves as a lancet to penetrate both the skin and the vein. The needle should be centered within the vein, and care should be taken not to go through the vein, i.e., out the other side.

(Continues)

GUIDELINES FOR ADMINISTERING MEDICATION

The veterinary technician must become competent in the delivery of medications as prescribed. It is important for the patient to receive the correct medication, by the correct route, and at the correct time. Routes of drug administration include injection, both **intramuscular** (**IM**) and **subcutaneous** (**SQ**), and some may be administered through a port on the intravenous (IV) line when a catheter is in place. Tablets and capsules are administered PO, by mouth. There may be circumstances in which a patient has difficulty swallowing or is "difficult to pill." In these situations, the tablet can be ground with a clean mortar and pestle (free of other drug residue) and a capsule can be separated. Both are then mixed with a small amount of water and administered by mouth with a syringe.

Prior to administering any medication it is very important to ensure the following:

- Correct patient: Reaffirm the identity from the cage card, patient file, and identifying paper collar placed on the patient when admitted.
- Correct drug: Read the label carefully three times—once when you remove it from the drug cupboard, again when you prepare the dose, and again prior to administering it to the patient.
- Correct dose: Make sure the dose is exactly as prescribed by the veterinarian and is measured accurately.
- Correct route: Administer drugs orally or by various injection sites.
- Correct time: Medications and treatments are most effective when administered in a timely manner. Different medications are given at different times and frequencies to obtain the maximum benefit.
- Understand how the prescription/treatment is written and commonly used abbreviations (**Table 1-1**).

DETERMINING THE CORRECT MEDICATION AND THE CORRECT DOSE

Occasionally, the veterinarian may write an order that simply states "3 mg amikacin SQ, bid." This requires translating and calculating. Translation: "Give 3 milligrams of amikacin, subcutaneously (by injection) twice a day." Twice a day is every 12 hours—for example, 6:00 a.m. and 6:00 p.m. (Easy).

In the drug cupboard, there are two bottles of AmiglydeV® but nothing labeled amikacin. Amiglyde V® sounds close, but is it correct? Never assume that "close enough" is the same. *Read every label carefully!* In this example, the label also reads "amikacin 50 mg/ml." (Amiglyde V® is the trade name for amikacin). Problem: The order is for 3 mg but the drug is labeled 50 mg/ml. How much of this medication should be given to the patient? (Not always so easy). One simple formula that can untangle the math is to remember is this:

$$\text{WT} \times \text{DOSE} \div \text{CONCENTRATION}$$

WT is the weight of the patient in *kilograms*. (Almost all medications are formulated by kilogram weight).

× (multiply) the dose, 3 mg

Divide by the concentration of the drug, 5C mg/ml. The patient's weight is listed as 16.5 pounds. To obtain the kilogram weight, divide the pound weight by 2.2 (1 kilogram = 2.2 pounds.) The patient's weight in kilograms is 7.5 kg. Multiply the kilogram weight (7.5) by the dose, 3 mg which equals 22.5.

Divide 22.5 by the concentration of the drug, 50 mg/ml. In other words, in each ml there is 50 mg of the drug, this is the concentration. 22.5 divided by 50 = 0.45 ml. This is the amount of this specific drug with this concentration that should be administered to the patient.

TABLE 1-1 Commonly Used Abbreviations for Administering Medications.

FREQUENCY	
sid	once a day
bid	twice a day
tid	three times a day
qid	four times a day
q	every
h	hour (q3h, for example, is every 3 hours)
DOSE	
ml	milliter (one ml and 1 cc are the same amount)
mg	milligram
g	gram
gtt	drop
L	liter
ROUTE OF ADMINISTRATION	
SQ	subcutaneous
IM	intramuscular
IV	intravenous
IP	intraperitoneal
OD	right eye
OS	left eye
OU	both eyes
AD	right ear
AS	left ear
AU	both ears
PO	by mouth
NPO	nothing by mouth, *nil per os*
TESTS AND PROCEDURES	
sx	surgery
CBC/chem	complete blood count and biochemistry panel
U/A	urinalysis

ADMINISTRATION OF ORAL MEDICATIONS

Many medications are formulated into capsules or pills that are given **orally**, straight into the mouth. As with all medications, check to be sure that it is the correct patient, the correct drug, the correct dose, and the correct time. The technique for administering medication orally should be mastered for both dogs and cats, as many oral medications are dispensed for further treatment at home. The veterinary technician is often called upon to demonstrate the easiest way for owners to give the medication to their pets. Sometimes it is recommended to hide the tablet in a small amount of food or a special treat. Dogs and cats are adept at eating every tasty morsel and "spitting out the pill." The only sure way of delivering the dose is to place it directly into the patient's mouth and ensure that it has been swallowed.

PURPOSE
- To administer medication orally

COMPLICATIONS
- Patient's inability to swallow capsule or tablet because of injury to the oral cavity
- Aggressive patient
- Scratches to the soft palate

EQUIPMENT
- Tablet or capsule to be administered
- 3–15 ml syringe without needle
- Small basin of drinking water

ADMINISTRATION of Oral Medication to a Dog

TECHNICAL ACTION	RATIONALE/AMPLIFICATION
1. Grasp the top of the muzzle and tip the head slightly back using your fingers and thumb, and gently press the lips against the molar teeth.	1a. Only a slight amount of pressure is required.
2. Use your other hand, holding the tablet, to open the bottom jaw with the middle finger, and place the tablet at the base of the tongue, near the throat.	
3. Maintain the hold on the dog's head, keeping it slightly elevated, remove fingers and hold the dog's mouth closed.	3a. Make sure the tongue is clear of the teeth before closing.
4. Encourage swallowing by stroking the dog's throat or by gently blowing into the nose.	

(Continues)

TECHNICAL ACTION	RATIONALE/AMPLIFICATION
5. Always inspect the inside of the mouth to make sure the pill has been swallowed and not tucked into a cheek fold or at the side of the tongue.	5a. It can help to follow up with 3 to 10 cc of water to encourage a deep swallow. The volume of water depends upon the size of the patient. The water should be delivered with the syringe into the side of the mouth and allow to patient to swallow. Do not attempt to force the water down the patient's throat.

ADMINISTRATION of Oral Medication to a Cat

TECHNICAL ACTION	RATIONALE/AMPLIFICATION
1. Hold the cat's head by placing one hand over the top of patient's head, grasping the upper jaw with the head tipped slightly back (this usually will cause the lower jaw to drop open).	1a. Some cats resist having their mouth opened and respond with aggression. It may be necessary to wrap the cat in a towel to prevent the cat from scratching with front or back claws. Care must always be taken to prevent a cat bite.
	1b. *Cat bites can be severe puncture wounds that quickly become infected. If bitten, seek medical treatment immediately.*
2. If, as above, the lower jaw does not relax and open, use the middle finger of the other hand to slightly pry it open.	
3. Hold the pill between the thumb and forefinger and place it in the center of the tongue toward the back of the mouth/throat.	3a. If the cat is aggressive, a pilling devices can be used—a plastic rod with a soft-tipped cup to hold the tablet and a plunger for administration. Be careful not to insert the "piller" too deep into the cat's throat or to damage the tongue or soft palate. If using this device, an additional restrainer is usually required.
4. Withdraw fingers and quickly close the mouth, tipping the head slightly downward to encourage the swallowing reflex.	
5. Observe the cat to ensure that swallowing has occurred and the pill has not been spat out or retained in the oral cavity.	5a. As with the dog, a few cc of plain water may encourage a deep swallow.

✳

ADMINISTRATION OF AN ORAL LIQUID

Oral medications are available in capsules, tablets, or liquid form. Many are now formulated with flavorings that are more acceptable to patients. If the patient has difficulty swallowing, tablets can be crushed and capsule contents can be emptied and dissolved in water. The solution can then be administered with a syringe. To do this, ensure that the entire volume is drawn into the syringe, tilt the head back slightly, and in the pocket of the lips, place the syringe between the lips and the rear molars. Deliver the contents of the syringe in small increments, allowing the patient to swallow. Never "slam" the contents directly into the throat, which could cause **aspiration**, or inhalation into the lungs.

ADMINISTRATION OF TOPICAL MEDICATIONS

Topical medications are those that are applied to the skin. They may include flea and tick preventatives or ointments and salves to treat skin abrasions and wounds. For flea and tick applications, carefully follow the manufacturer's directions. Always wear disposable gloves and dispose of the empty applicator appropriately by enclosing it in the gloves when they are removed.

If using a topical medication for an open wound or abrasion, the area should be thoroughly cleaned and **debrided**, that is, removing dead skin, forming scabs and dried **exudate**.

If necessary, the hair from around the site should be shaved. Clean the wound with a surgical scrub and rinse thoroughly with 0.9% saline, and allow the skin to dry prior to applying the medication. Do not use alcohol on an open wound.

ADMINISTRATION OF AURAL MEDICATION

PURPOSE

- To administer medication to the ear

COMPLICATIONS

- Inadequate restraint
- Use of a partial tube from another patient
- Introducing infection or contaminates from partially used opened tube

EQUIPMENT

- Restrainer
- Ear cleaning supplies
- New tube of prescribed medication

PROCEDURE for Applying Medication to Ear

TECHNICAL ACTION	RATIONALE/AMPLIFICATION

1. Thoroughly clean the affected ear (refer to grooming procedures).

2. Place the medication tube or dispensing tip within the ear, being careful not to advance the tip of the applicator into the ear canal (**Figure 1-60**).

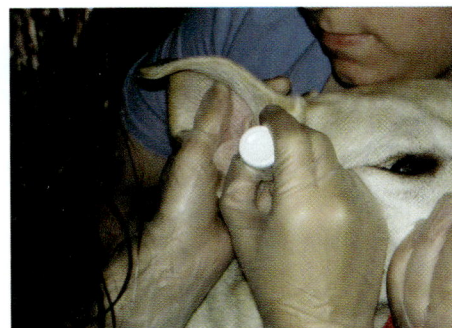

▲ **FIGURE 1-60** Restraint and positioning for administration of medication into the ear.

(Continues)

TECHNICAL ACTION	RATIONALE/AMPLIFICATION
3. Administer the correct amount of medication.	3a. **Aural** (ear) medication is often prescribed by drops; be sure to count the number of drops so the correct amount is administered.
	3b. If the medication is an ointment and comes in a tube, clean the end of the tube and label it with the patient information. To avoid spreading infection, never use an opened tube on different patients. Frequently, these tubes are dispensed for further treatment at home. The tube can be placed in a clear bag or a dispensing vial with the prescription and dose written on the label.
4. Remove the dropper or dispenser from the ear.	
5. Massage the outside base of the ear.	
6. Repeat the procedure with the other ear if so prescribed.	

ADMINISTRATION OF OCULAR MEDICATION

PURPOSE
- To administer medication to the eye

COMPLICATIONS
- Re-use of non-sterile opened vial or tube
- Introducing infection
- Scratches to the cornea
- Inadequate restraint

EQUIPMENT
- Restrainer
- Sterile unopened tube or vial of drops as prescribed

ADMINISTERING Eye Medication

TECHNICAL ACTION	RATIONALE/AMPLIFICATION
1. Hold the eye lids open and tilt the head slightly back. The hand holding the medication should be stabilized and resting on the animal's head.	
2. Administer drops by holding the dropper slightly above the open eye. Administer an ointment by placing a thin layer of the medication on the on the lower eyelid (**Figure 1-61**).	2a. Avoid touching applicator tip to the eye to prevent damage to the eye. Stabilizing the hand holding the medication by resting it on the patient's head helps prevent direct contact with the eye.

(Continues)

TECHNICAL ACTION	RATIONALE/AMPLIFICATION

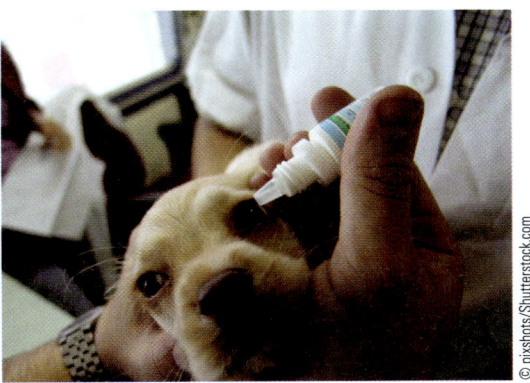

▲ **FIGURE 1-61** Restraint and positioning for administration of medication to the eye.

3. Allow the patient to blink several times to distribute the medication across the entire eye.

3a. **Ocular** (eye) medications are sterile; do not contaminate the tip when handling. Eye medications, if not for single use, should be packaged and dispensed in the same manner as aural medications or the open tube disposed of correctly.

ADMINISTERING MEDICATION BY INJECTION

The most frequently used injection routes are **subcutaneous (SQ)** (under the skin) and **intramuscular (IM)**. Refer to **Figure 1-62**, illustrating the delivery of various injection sites and the approach.

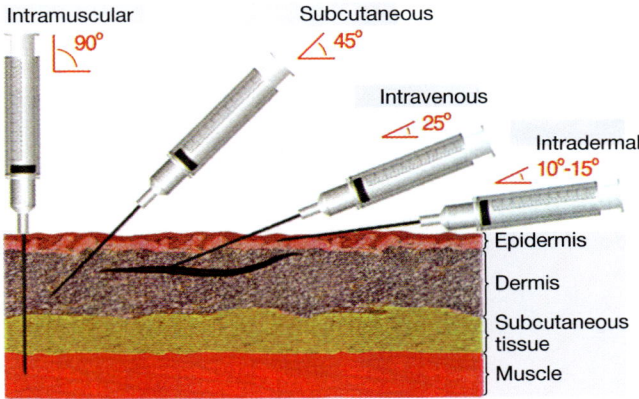

▲ **FIGURE 1-62** The correct angle of approach for delivering medication by injections.

PURPOSE OF SQ INJECTIONS
- To deliver the prescribed dose of medication or vaccine under the skin

PURPOSE OF IM INJECTIONS
- To deliver a medication into the muscle tissue

COMPLICATIONS OF SQ INJECTIONS
- Delivering the injection into the layers of skin
- Delivering the injection "intra-fur"— that is, pushing the needle too far into the skin and out the other side. Not only does this waste the drug or vaccine, but the patient will also require another stick of the needle.

COMPLICATIONS OF IM INJECTIONS
- Restraint of patient, as IM injections are painful
- Inadvertent injection into bone, tendon, or sciatic nerve

EQUIPMENT FOR INJECTIONS
- Appropriate-sized syringe with attached sterile needle.
- Alcohol wipes, either cotton balls or gauze
- Medication or vaccine to be delivered

RABIES Vaccines

Rabies vaccinations are required by law, and in most states only a veterinarian is permitted to vaccinate a patient against rabies. The veterinary technician may prepare the rabies vaccine and deliver any other vaccines or medications as prescribed by the veterinarian, but not rabies.

PROCEDURE for Administration of an SQ Injection

SQ injections are less painful and are delivered under the loose skin at the nape of the neck and between the shoulder blades. It is important that an SQ injection is not injected into the layers of the skin, **intradermal (ID),** as this affects the time for drug delivery and may cause localized inflammation.

TECHNICAL ACTION	RATIONALE/AMPLIFICATION
1. Remove the cap from the needle, and with the needle attached, draw the correct amount of drug into the appropriate-sized syringe.	1a. Needles are gauged according to the size or lumen of the needle. For example, an 18 gauge needle is larger than a 27-gauge needle. For most SQ injections, choose a 25- to 22-gauge needle, depending upon the size of the patient. Cats are more tolerant of the injection when a smaller 25 gauge needle is used. The syringe should be chosen according to the volume that has to be administered, and usually a 3 ml syringe is appropriate. All needles can be changed and attached to any size syringe.
2. Check the volume that has been drawn into the syringe, and tap the side of the syringe to remove any air bubbles.	

(Continues)

TECHNICAL ACTION	RATIONALE/AMPLIFICATION

3. Replace the cap on the needle to keep it sterile.

4. Lift the loose skin between the shoulder blades so it forms a small tent, and wipe the injection site with alcohol using either a cotton ball or gauze pad.

5. After removing the cap from the needle, insert the needle under the skin at the base of the tented area (**Figure 1-63**).

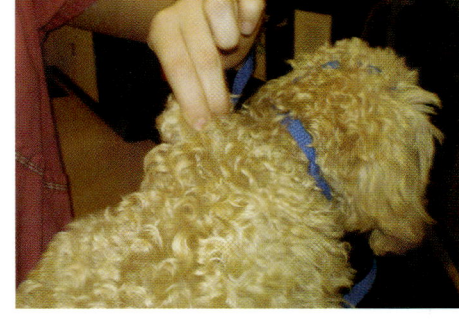

▲ **FIGURE 1-63** To administer a subcutaneous injection, lift the loose skin of the shoulders to form a tent, and inject the medication just under the lifted skin.

6. Pull slightly back on the syringe plunger to confirm that there is no blood in the hub of the syringe. If no blood appears, deliver the entire contents of the syringe in one smooth motion.

6a. If blood appears in the hub, do not inject, as the needle may have entered a small vein. Withdraw and insert into another site. When doing this, change to a new needle to avoid contamination.

7. Withdraw the needle and gently massage the injection site.

8. Dispose of used needle and syringe correctly.

8a. Remove all used needles and place them directly into the sharps container. Syringes are designed to be disposable and should not be used again.

PROCEDURE for Delivering an Injection IM

IM injections are more painful because the needle is inserted and the medication delivered directly into a muscle mass. Instructions for the correct route must be followed: Drugs that are formulated for SQ injection should never be administered IM, nor should IM drugs ever be delivered SQ.

TECHNICAL ACTION	RATIONALE/AMPLIFICATION

1. Determine the correct location for the injection site: Frequently used sites are the quadriceps of the rear limb, the muscle group either side of the lumbar spine (epaxial muscles), and occassionally, the triceps muscles of a fore-limb. Select the appropriate-sized needle and syringe.

1a. When using a muscle group in the thigh, an injection delivered incorrectly can injure the sciatic nerve. It is preferable to choose another location; however, if these muscles are used, the needle should be directed **caudally** (toward the tail) with a 45° angle to avoid potential damage to the sciatic nerve. Injecting into the sciatic nerve can cause permanent damage and, potentially, paralysis of the affected leg.

(Continues)

TECHNICAL ACTION	RATIONALE/AMPLIFICATION
2. Determine preferred injection site with the restrainer as the technique of restraint will depend on injection site.	
3. Follow steps 1–3 for SQ injection.	
4. With the patient standing, have an assistant restrain the head and body.	
5. Locate the chosen muscle group by palpation.	
6. Swab the injection site with alcohol.	
7. Quickly insert the needle and aspirate the syringe to make sure the needle hasn't entered a vein. (If blood flashes in the hub of the needle, the needle will have to be withdrawn, changed, and placed in another location).	
8. Inject the medication slowly.	**8a.** Slow to moderate injection of the medication is less painful.
9. Withdraw the needle and massage the injection site.	

ADMINISTRATION OF MEDICATION INTRANASALLY OR INTRADERMALLY

Occasionally, medication or a vaccine (Bordatella) may be delivered **intranasally (IN)**. Many respiratory medications are delivered directly into the nares for rapid assimilation. Always use a new syringe with the needle removed: *Never* attempt to deliver the medication with a needle still attached. The patient may be sitting or in sternal recumbency. With one hand restraining the top of the muzzle, tip the patient's head back so the nose is slightly elevated, deliver the medication directly into the nostrils (**Figure 1-64**). Patients frequently attempt to sneeze and clear the medication, so it is important to hold the head slightly elevated long enough to allow the medication to permeate the nasal mucosa.

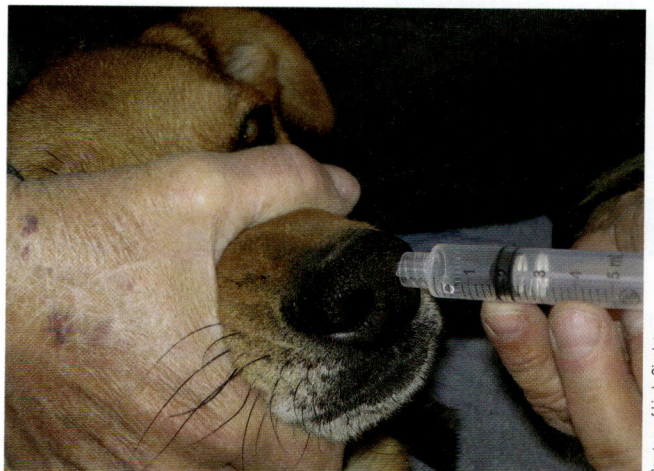

Courtesy of Linda Singleton

▲ **FIGURE 1-64** When administering an intranasal medication, the head should be held slightly back to allow the medication to enter the nasal mucosa.

Intradermal injections (ID) usually are given when testing for allergies and skin diseases. This method is also used to deliver local anesthesia to suture a small wound when general anesthesia is not required. The area of the injection should be shaved with a # 40 blade and cleaned with alcohol or another antiseptic solution. The needle is inserted at a slight angle and between the layers of skin. When injected correctly, a small bleb should be apparent between the skin layers.

ADMINISTRATION OF FLUID THERAPY

Fluid therapy is a frequent supportive and, in some cases, life-saving treatment administered to patients of all species. Fluid support, or fluid therapy, is administered to all critical care patients, to trauma patients to help alleviate shock and compensate for blood loss, to support surgical patients, and to all dehydrated patients to restore and maintain **hydration status**.

Hydration status is an estimate of the amount of *dehydration*, the fluid deficit, and is measured as a percentage (see **Table 1-2**). This determines not only the need for fluid therapy but also the volume of replacement or maintenance fluids needed. A complete blood and chemistry profile can determine the type of fluids that are administered. Evaluation of hydration status is an important part of the physical examination; refer to it for review, skin tenting and significance of the CRT.

TABLE 1-2 Estimating the Percentage of Dehydration.

5%:	undetectable
5–6%:	some elasticity of skin is lost
6–8%:	skin tenting is slower to return to normal
	slower CRT (normal is 2 seconds or less)
	eyes may be sunken into the orbits
	mucous membranes may be dry
10–12%:	skin remains tented
	CRT is prolonged
	eyes sunken into the orbits
	dry mucous membranes
	possible shock
12–15%:	shock (rapid HR, weak pulse)
	death

Approximately 60% of the body is composed of water, or body fluids. Body fluids are **intracellular**, within the cells; **interstitial**, between the cells; and **intravascular**, within the entire circulatory system. Body fluids circulate constantly, delivering oxygen and nutrients to maintain cellular function. Fluids also deliver **electrolytes**, chemical substances essential to maintaining all body functions. Fluids also remove waste products, and the constant movement of fluids in and out of the cells keeps the body consistently maintained, functioning, and adapting, in a state of **homeostasis**. If homeostasis is disrupted as the result of an imbalance of electrolytes or volume of circulating fluids, fluid therapy is essential to correct these imbalances and return to homeostasis.

ELECTROLYTES

Electrolytes are chemicals that when dissolved in water become electrically charged particles (ions). Ions are either positively charged (+) or negatively charged (–). Positively charged ions in body fluids are:

- Sodium (Na+)
- Potassium (K+)
- Calcium (CA+)
- Magnesium (Mg+)

Negatively charged ions in body fluids are:

- Chloride (CL–)
- Bicarbonate (HCO)
- Phosphate (PO)

The type of fluid and the amount prescribed by the veterinarian is specific to the need, or deficit, determined by the estimated percentage of dehydration, a biochemistry profile, and whether fluids replace or supplement these deficits. The veterinary technician must, as with all other medications and therapies, recognize and confirm the correct type of fluid that has been prescribed for a specific patient. They are *not* interchangeable.

COMMONLY USED FLUIDS

All solutions are formulated with sterile water

0.9% Saline solution: sodium chloride in the same amount as blood serum

Lactated Ringer's Solution (LRS): electrolytes in the same concentration as blood; sodium, potassium, calcium, chloride, and bicarbonate. The bicarbonate is the lactate (body salts).

Ringer's Solution: sodium chloride, calcium chloride

Fluids are described as being **isotonic**, **hypotonic**, and **hypertonic**. Isotonic fluids are chemically the same as normal body fluids and cause no change to the cell. Hypotonic solutions cause cells to enlarge or expand, and hypertonic solutions cause cells to shrink. The effect of each type of fluid, called **tonicity**, determines how fluids are distributed within the body.

Frequently, tonicity is determined by the percentage of electrolytes already present in the type of fluid that is administered or by adding a prescribed amount of another substance to an isotonic solution.

Fluids can be delivered by various routes, depending on the volume required. Subcutaneous delivery is frequently used, and the sites for injection vary depending on skin elasticity reflected by hydration status. Only isotonic fluids should be delivered SQ to avoid skin irritation, damage to the subcutaneous tissue, and, in some instances, skin sloughing.

Intravenous delivery requires catheter placement. Fluids are distributed more quickly and the total volume is calculated as drops per minute. IV catheters should be monitored for patency and signs of **phlebitis**, an inflammation of the vein. IV catheters should not be left in place longer than 48 hours. A catheter can be placed in the cephalic vein, lateral saphenous in the dog, and medial saphenous in the cat, or in the jugular vein.

Intraosseous fluids are placed directly into the intramedullary space in either the head of the femur, the cranial aspect of the humerus, or the wing of the ilium. Fluids can be delivered with a needle and syringe or with catheter placement. IO delivery requires local or general anesthesia, sterile preparation of the site, and there is a risk of introducing infection to the bone.

Less frequently, fluids can be delivered directly into the intraperitonal space. Rarely used in mammals, it is a common approach to rehydrating a reptile patient.

In addition, fluids can be delivered orally provided that the patient is able to swallow and has no history of vomiting. Using the same method as in giving oral medications, fluids are delivered with a syringe. It is extremely important to deliver the fluids in small increments, allowing the patient to swallow. This is to prevent aspiration and potentially drowning the patient if fluids are inhaled into the lungs.

As with all medications, the veterinarian determines the type of fluid to be administered, the volume to be administered, and any substance added to the prescribed type of fluid. The veterinary technician's role is not only to deliver SQ fluids but also to set up and maintain fluid bags and the delivery system, to calculate and monitor delivery rate, to maintain the patency of a fluid delivery system, to deliver and become proficient in IV catheter placement, and to maintain catheter patency.

PURPOSE OF FLUID THERAPY

- To restore hydration status
- To correct electrolyte imbalances
- To maintain organ perfusion and function during a surgical procedure
- To maintain hydration status.

COMPLICATIONS

- Puncture of fluid bag

EQUIPMENT FOR SQ DELIVERY OF FLUIDS (FIGURE 1-65)

- Sterile needle (18–22 gauge)
- Prescribed bag of fluids to be delivered
- IV pole
- IV set
- Calculated amount to be delivered
- Sharpie type black pen
- Strip of white tape

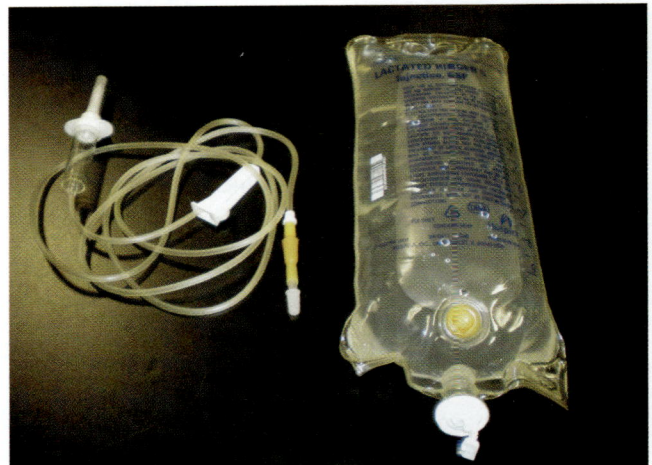

▲ **FIGURE 1-65** A liter bag of IV fluid and the infusion or administration set.

SIZE OF IV BAGS

Standard bags of fluids contain 1,000 ml—1 liter. There are also 500 ml bags, commonly referred to as "pedibags" (pediatric), used when smaller amounts of fluids are required. Volume on both is marked in milliliter increments on the side of bag.

PROCEDURE for Preparing Fluid Bag and Delivery System

TECHNICAL ACTION	RATIONALE/AMPLIFICATION
1. Confirm fluid type before opening.	1a. As with all medications and treatment, confirm that it is the correct type of fluid required.
2. Remove and discard the heavy outer covering of the bag.	
3. Open a new infusion set and become familiar with the parts.	3a. Refer to the packaging of the delivery system, although there several manufacturers all function in the same way. Often, there is a diagram of the parts and instructions on the packaging.
	3b. Check the type of drip sets normally used in the clinic. The **macrodrip** delivery system is most commonly used. Macrodrip systems are designed with different-sized drops to equal 1 ml volume. For example, a macrodrip system that is rated 10 means that 10 drops equal 1 ml. of fluid; a macrodrip system rated 15 means that 15 drops equal 1 ml and a system rated as 20 means that 20 drops equals 1 ml. The size of the drops released determines the number of drops per milliliter.

(Continues)

TECHNICAL ACTION	RATIONALE/AMPLIFICATION
	Microdrip sets are used only in very small animals or neonates when small amounts of fluids are prescribed over a set amount of time. With all microdrip delivery sets, there are 60 drops in 1 ml of fluid. In other words, it would take 1 hour to deliver 1 ml of fluid at a drip rate of one drop/second. Because of this and the difficulty of catheter placement and maintenance, the use of microdrip sets for fluid therapy is limited.
4. Close both clamps: One is a slide clamp; the other is a regulating clamp.	4a. Do not stretch the tubing. The flow control clamps may be compromised and there will be a loss of flow control effectiveness.
5. Remove the cover (cap) from the spike end of the infusion set and discard.	
6. Remove the spike port cover on the bottom of the bag of fluids and discard.	
7. Push spike end of the infusion set firmly into the port on the fluid bag.	7a. The spike on the infusion set must be pushed straight into the spike port to avoid puncturing the bag.
8. Hang the bag from the IV pole.	
9. Remove the protective cap from the other end of the infusion line and replace it with an 18–25 gauge needle (retain needle cap).	9a. It is important to keep the needle clean and replace the needle cap to lower the risk of bacterial contamination.
10. Open the slide clamp and fully open the wheel on the regulating clamp.	
11. Allow an open flow of the fluids so the line is full.	11a. By allowing free flow of the fluids throughout the entire line, it not only proves the patency of the line itself but also forces any air out of the system. When doing this, allow fluid to drain into the sink, not on the floor!
12. Close both clamps and re-cap the needle.	12a. Always close both clamps when not in use, to prevent accidental loss of fluids if the needle becomes dislodged.
13. Drape the fluid line over the pole and it is ready for use.	13a. Keep the line neat and out of the way to avoid damage to the line and bag.

PROCEDURE for Delivering SQ Fluids

TECHNICAL ACTION	RATIONALE/AMPLIFICATION
1. Confirm patient and type of fluids ordered.	
2. Obtain accurate weight of patient in kilograms.	2a. 3a. This is determined by the estimated percentage of dehydration (refer to Table 1-2 for clarification) multiplied by the body weight in kilograms to determine the estimated fluid deficit in milliliters.
3. Determine the total amount of fluids to be administered over a 24-hour period.	

(Continues)

TECHNICAL ACTION	RATIONALE/AMPLIFICATION
4. With the Sharpie® pen, mark the piece of tape with patient information, date, and time, then place the tape on the side of the bag at the millimeter mark to reflect the stopping point, or when the volume has been delivered.	4a. This provides a visual, as well as confirming that the correct volume of fluids has been delivered.
5. Prepare the area where fluids will be administered (table, towel, and equipment close to hand and in one location).	5a. All equipment should be at hand and ready for the patient.
6. Replace the needle at the end of the infusion line with an 18–25 gauge needle.	6a. The original needle placed at the end of the infusion line is basically a plug. Re-cap this needle and dispose of it in the sharps container. By getting in the habit of always changing the needle it avoids potential cross contamination between patients.
7. Place patient on table, and reassure while manually restraining with one hand across the shoulders.	7a. Dogs and cats rarely object to the administration of SQ fluids and an additional restrainer is usually not required. Cats usually settle themselves in sternal recumbency. Dogs may either be sitting or in sternal recumbency.
8. Remove cap from new needle.	
9. Tent the skin over the shoulder blades, insert needle, and open both clamps. (Refer to Figure 1-54.) The placement and procedure are similar to administering a SQ injection.	9a. This is the same location and procedure used for other subcutaneous injections.
10. Release tented skin, and maintain one hand over the puncture site.	10a. The tented skin is released to allow a freer flow of fluids. The hand over the puncture site helps stabilize the needle.
11. Monitor volume delivered into each injection site.	11a. No more than 5 to 10 ml should be delivered into one site. Volumes greater than this cause some discomfort to the patient and, because of limited subcutaneous space, some of the fluids may leak out. Adjust the position of the needle or withdraw and chose another site. Multiple sites should be used rather than delivering the entire volume into one site.
12. When total volume has been delivered, close both clamps and remove needle.	12a. Re-cap the needle and drape the line back over the infusion pole. Never leave a needle un-capped to avoid an accidental stick and to prevent bacteria from entering the fluid line.
13. Return patient to cage and record the procedure in the patient record.	

For example, a cat weighs 8 pounds. First, the pound weight has to be converted to kilograms. Divide 8 pounds by 2.2 to obtain the weight in kilograms (approximately 3.6 kilograms). Next, the cat's estimated percentage of dehydration is 5%. Express the percentage as a decimal (.05), and multiply this by the body weight in kilograms (180); 180 represents the fluid deficit *per kilo* weight of the patient.

A guideline used to calculate the total volume required is 40-60 ml per kilogram per day (40–60 ml/kg/day).

For example, use the median average of 40–60 ml/kg/day, which is 50 ml/kg/day.

Multiply 50 × 180 to equal a total volume of 900 ml/day.

Subcutaneous fluids are absorbed slowly, and it is better to divide the total volume per day into three or four delivery sessions. To determine the volume by the number of sessions, divide 900 by 3 or by 4 (three sessions or four sessions). Always divide the total volume per day by the number of hours available.

GUIDELINES FOR CATHETER PLACEMENT

There are several types of intravenous catheters, but one of the most frequently used is the **over-the-needle catheter**. This is exactly as the name implies: The catheter, which is placed and remains within the vein, is designed with a needle inside the catheter. The needle performs the puncture and guides the catheter into the vein. Once placed, the needle is withdrawn, leaving the catheter within the vein.

Catheters, like needles, are available in different gauges and lengths. Always consider the length and gauge of the catheter relative to patient size, the site (vein) where it is to be placed, and choose the smallest gauge catheter but one large enough for the required delivery rate. It must be remembered, too, that the smaller the gauge of the catheter the greater the potential for occlusion and the longer the catheter is the more stable it is within the vein.

- For dogs and catheter placement in the cephalic or lateral saphenous vein, a 1½ inch, 18- to 20-gauge catheter is a good choice.
- For cats, and placement in either the cephalic or medial saphenous vein, the 1½ inch 20 gauge catheter is more suitable (**Figure 1-66**).

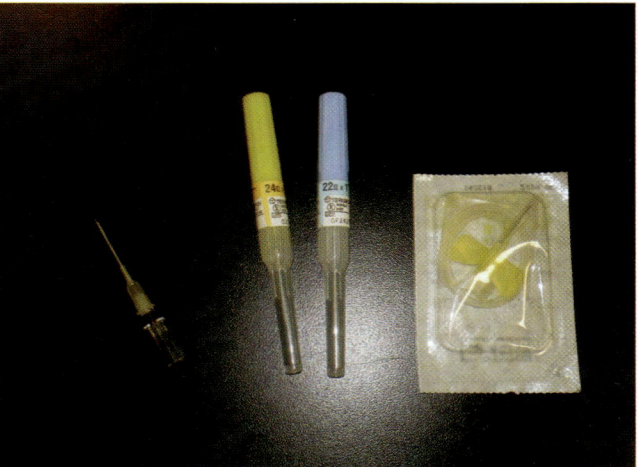

▲ **FIGURE 1-66** A selection of intravenous catheters.

PURPOSE

- To gain venous access for medication administration

COMPLICATIONS

- Introduction of bacteria
- Non-patent catheter
- Phlebitis
- Swelling at catheter placement site
- Patient chewing on delivery lines or catheter
- Occlusion (blood clot)
- Laceration of vein
- Overhydration

EQUIPMENT

- Clipper with # 40 blade attached
- Disposable gloves
- Sterile gloves
- Gauze sponges soaked in providone iodine or chlorhexedine solution
- Cotton balls or gauze sponges soaked in 70% isopropol alcohol
- IV catheter
- Injection cap
- Primary IV set
- Antibiotic ointment
- Half-inch-wide porous tape strips (2)
- One-inch-wide porous (white) tape strip
- Roll of cotton (Kling® or similar) bandaging material
- Roll of self-adhesive bandaging material (Vetwrap® or similar)
- Heparin flush

CATHETER Placement

TECHNICAL ACTION	RATIONALE/AMPLIFICATION
1. Gather together all needed items.	1a. Tear all tape strips long enough to wrap around the limb twice and stick the ends to the procedure table where they can be quickly reached.
2. Confirm patient.	
3. Enlist a restrainer for the procedure.	3a. Restrain the patient to prevent laceration of the vein.
4. Place patient on table and assess vein access.	4a. The health status of the patient and previous blood draws will affect the viability of different veins. Choose the one that is the most easily accessed.
5. With a # 40 blade, shave the area approximately twice the length of the catheter and two blade widths either side of the vein.	5a. The clipped area provides for easier visibility and aseptic preparation of the catheter placement site.
	5b. Always attempt placement at the most distal end of the prepared site. This allows a second attempt (if necessary) above (proximal to) the first site.

(Continues)

TECHNICAL ACTION	RATIONALE/AMPLIFICATION
	5c. By placing the catheter more distal, it avoids the chance that when the patient flexes the limb, the end of the catheter can become occluded at the elbow. Use a splint to keep the limb straight if necessary.
6. Don disposable gloves and perform a sterile preparation of the shaved area using aseptic technique.	6a. Clean the site, using the aseptic technique to reduce the chance of inadvertently introducing bacteria, dander, and small particles of hair.
	6b. Use the gauze pads soaked in either providine iodine or chlorhexadine solution to clean the entire shaved area. Use the pads in a circular, outward motion from the targeted puncture site. To avoid recontamination, do not go back over the site with the same gauze pad, but discard and use a new one. Continue until there is no debris or dirt on the gauze pads. Follow up using the gauze pads soaked in isopropal alcohol.
7. The restrainer should position the patient and occlude the vein, using the same technique as for a blood draw.	
8. Clear the area of used gauze pads, remove disposable gloves, and replace with sterile gloves.	8a. Change to sterile gloves to further reduce the risk of contamination of the catheter placement site.
9. Hold the catheter, bevel up, and enter the vein at a 15–30° angle.	9a. The bevel of the needle should always be up. It avoids trauma to the vein and permits a freer blood flow.
10. Confirm needle/catheter placement into the vein, and decrease the angle of the needle and catheter.	10a. A flash of blood will appear in the hub of the needle confirming vein access.
	10b. Decreasing the angle of the needle makes catheter placement into the vein easier and smoother.
11. Advance the needle into the vein and smoothly slide the catheter into the vein as far as the hub, and withdraw the needle.	11a. Inserting the catheter to the hub makes the catheter more stable. The end of the catheter should not be left protruding. This also removes the risk of pulling out the catheter when the needle is removed. To help prevent this, stabilize the vein with your thumb when withdrawing the needle.
12. Hold the catheter in place and attach the injection cap or attach directly to a primed infusion line.	12a. The injection cap packaging should be opened and ready for use. While maintaining stabilization of the vein, insert the cap into the open end of the catheter. Blood will flow from the catheter until capped or connected to a primed IV line.
	12b. An advantage of connecting straight to the delivery system allowing a small amount of fluids to flow through the catheter helps to stabilize the vein while bandaging the catheter placement site.
	12c. Remove all traces of blood from the placement site. Blood is a fertile ground for bacterial growth.
13. Flush the catheter with a small amount of heparanized saline to ensure patency.	13a. Prior to bandaging, flush the catheter with ½ to 1 ml of heparinized saline to ensure patency.

(Continues)

TECHNICAL ACTION	RATIONALE/AMPLIFICATION
	13b. The heparin/saline flush should be drawn into a syringe from a stock bottle of "flush" or for smaller amounts, heparin is added directly to the syringe. The standard ratio of heparin to 0.9% saline is 500 IU of heparin to 250 ml of 0.9% sterile saline. To prepare a single use syringe, add 2 (two) IU of heparin to 1 ml of 0.9% sterile saline.
	13c. 0.9 % sterile saline is also available in rubber stopped bottles of 250 and 500 ml. If using a prepared heparin flush from a ready-prepared bottle, always swab the rubber cap with alcohol prior to inserting a needle and withdrawing the amount needed in a single syringe.
14. Cover the puncture site with a topical antibiotic, and bandage the catheter in place.	**14a.** Place a small amount of topical antibiotic ointment on a sterile gauze pad and lay it directly over the puncture site. This acts as another barrier against potential infection.
	14b. Take one of the narrow strips of tape, sticky side up, and slide approximately ½ inch of the tape under the catheter port hub. Fold a small tab at the long end of the tape for easier removal. Wrap the longer end completely around the limb and over the top of the catheter, covering the gauze pad. Be careful to keep the gauze pad smooth and flat with no wrinkles or folds. The tape should ***not*** be pulled tightly; it is used only to stabilize the catheter.
	14c. Place the second narrow strip of tape, sticky side up, under the distal end of the catheter cap so it wraps around the shaved area yet prevents border hair from touching the catheter.
	14d. Use the cotton roll for the first layer of bandaging, being careful to leave the catheter port exposed.
	14e. Next, wrap the site with gauze bandaging material to form the second layer.
	14f. Make a small loop with the IV tubing and lay it smoothly onto the gauze layer. Be sure that it is not twisted or kinked.
	14g. While holding the loop in place, apply the third and final layer of the bandage (Vetwrap® or similar self-adhesive bandaging material) over and under the loop so it remains visible. Forming this loop of IV tubing helps prevent dislodging the catheter if the line is tugged, caught up in a cage door or when moving the IV pole.
	14h. Secure the end of the third layer with a piece of wide white tape if necessary. (White tape should never be applied directly onto the skin.)
	14i. Check the IV port is clearly accessible and the bandage is not too tight. It should be observed regularly for any signs of swelling, heat, and discomfort to the patient.

(Continues)

TECHNICAL ACTION	RATIONALE/AMPLIFICATION
15. Determine that the flow rate of the fluids is correct, and adjust as necessary.	**15a.** Release the slide clamp and adjust the regulator clamp to achieve the correct flow rate. The drip chamber can be manually squeezed to start the drip. Allow the drip chamber to fill to no more than ½ of the capacity of the chamber. Always remember to mark the bag of fluids with the volume and time therapy begins.

FLOW RATE

There are two types of IV drip sets: A **macrodrip** set delivers 10, 15, or 20 drops/ml. The **microdrip** set delivers 60 drops/ml. The type of fluid delivery set is always marked on the packaging, and it is important to know which set is being used so as to most efficiently deliver the correct amount of fluids over the determined amount of time, the **flow rate**. The flow rate equals the volume in mls per hour or total volume per day.

Drops are counted as they fall into the drip chamber. The flow rate is adjusted with the regulator clamp. By turning the wheel upward, the drip rate increases as the regulator wheel opens the IV line. Turning the wheel downward narrows the line to decrease the number and frequency of drops. When the regulator wheel is all the way to the top of the clamp, the flow rate is "wide open." "Wide open" is often used as a directive to deliver the maximum amount of fluids in an emergency effort to help stabilize a patient, but it should never be left in this position for normal fluid rate delivery, at the risk of **overhydration**. Overhydration will cause serious complications and further compromise a patient. Signs of overhydration—delivering too much fluid volume—include restlessness, shivering, a decrease in heart rate, an increase in respiratory rate, and coughing. All patients receiving fluid therapy need to be monitored carefully, and the veterinary technician must always determine that the fluid delivery rate is accurate.

CALCULATING FLOW RATE

Once the total volume of fluid replacement is determined, the flow rate can be calculated. For example, the patient requires 2,150 ml over 12 hours.

Divide the ml amount (2,150 ml) by the number of hours (12) = 179. 1666 ml/hour. It is not possible to deliver this exact amount, and the number is round down to 179 ml/hour.

To determine the drops/sec, divide 179 ml/hour by 60 (number of minutes in an hour) = 2.98 drops/minutes.

As above, this exact amount cannot be delivered, and 2.98 is rounded up to 3 drops/minutes.

The total volume and type of fluids are determined by the veterinarian. It is determined by percentage of dehydration, the daily maintenance requirement, and to compensate for ongoing losses. Ongoing losses are an estimation of fluids lost due to urine output, vomiting, and diarrhea.

In summary, a patient's daily fluid requirement = fluid deficit + maintenance + ongoing loss

INFUSION PUMPS

Infusion pumps are designed to automatically deliver a pre-set volume and delivery rate for fluids (**Figure 1-67**). They must be carefully programmed or the purpose is defeated. Using an infusion pump does not mean "set and ignore." The IV line must be carefully threaded through the pump and accurately programmed for the delivery rate. Infusion pumps have an alarm that alerts personnel if the fluids are not flowing or when the fluid bag is empty. If the alarm is functioning, it may indicate that the IV line is occluded, the line needs to be flushed, or the tubing rearranged. The catheter also should be checked for patency and that the clamp controlling the flow is open. Depending on the specifics of the machine (pump), it may then have to be reprogrammed.

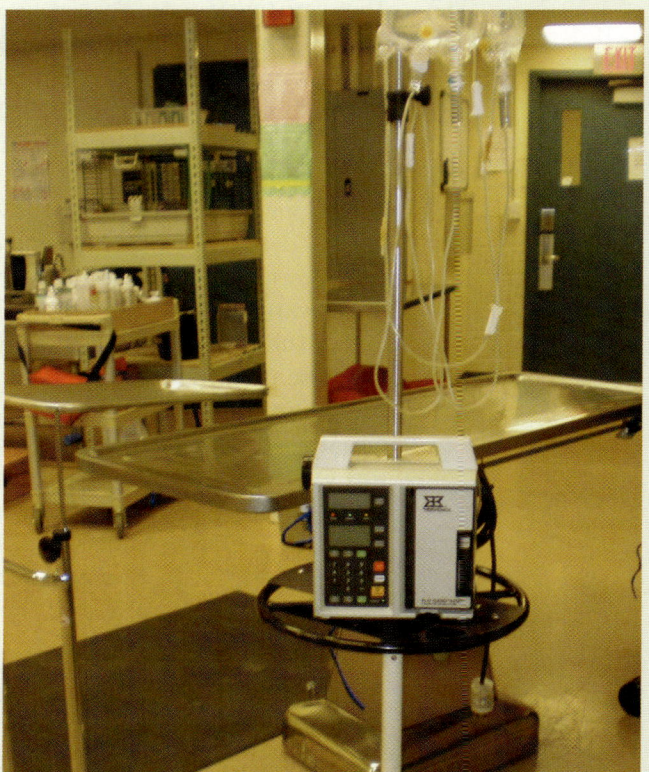

▲ **FIGURE 1-67** The IV infusion pump can be set to deliver a specific volume of fluids at a predetermined rate. Infusion pumps must be accurately programmed and monitored to ensure that they are functioning correctly.

Anesthesia

© Liliya Kulianionak/Shutterstock.com

OBJECTIVES

Upon completion of this unit, the reader should be able to:

▶ Understand basic principles of anesthesia
▶ Describe the importance of pre-anesthesia drugs and pain control
▶ Understand injectable anesthetics and their use
▶ Identify the components of the anesthesia machine and delivery system
▶ Perform pre-anesthesia function tests for the equipment
▶ Understand the stages of anesthesia
▶ Demonstrate use of the Anesthesia Record Form
▶ Monitor and record a patient's vital signs during anesthesia
▶ Induce, maintain, and recover a patient from anesthesia
▶ Recognize and respond to anesthetic emergencies

KEY TERMS

analgesic	endotracheal tube	peristalsis
anesthesia	flow meter	pop-off valve
anesthesia machine	flush valve	pre-anesthetics
Anesthesia Record Form	induction	pressure gauge
anticholinergic	inhalation anesthesia	recovery
breathing system	maintenance	vaporizer
	palpebral reflex	

ANESTHESIA

Delivering anesthesia to a patient is the most demanding procedure a veterinary technician will perform. It is the most critical area of knowledge that a technician must possess and, without overstating the importance, inducing, maintaining, and recovering patients, literally places the life of the patient in the hands of the anesthetist. Every heartbeat, every breath, every second of the patient's time during anesthesia and recovery depend on the knowledge, skills, and vigilance of the anesthetist.

Anesthesia literally means "without sensation." *General anesthesia* produces a loss of consciousness and a loss of pain sensation during a procedure. Anesthesia is also used as a method of restraint for a patient with a painful condition or severe injury during the initial examination and to restrain an otherwise unapproachable animal, for radiographic positioning, and in performing euthanasia.

Local anesthesia is administered to block nerves and alleviate sensation (pain) in a *localized* area, to suture a small wound, for example; there is no loss of consciousness.

What anesthesia does *not* do is prevent pain; unconsciousness does not mean the absence of pain, only absence of the perception of pain.

PRE-ANESTHETICS AND PAIN CONTROL

Patients undergoing surgical procedures are routinely given **pre-anesthetics**, drugs that alleviate pain before, during, and after surgery. **Analgesic** agents, pain-relievers, can also reduce the amount of anesthetic agent required to maintain a patient during a procedure. A term that has come into general use from human medicine is *preemptive analgesia*; to predict and prevent pain before it occurs. It is always easier to prevent pain than it is to alleviate it. Pain control has become an increasingly important aspect of successful veterinary anesthesia. Pain delays recovery, healing, and mobility.

Pre-anesthesia agents include sedatives or tranquilizers to relieve patient anxiety and cause muscle relaxation. Pre-anesthesia protocol may also include an **anticholinergic**, a drug that prevents bradycardia (decrease in heart rate), excessive salivation, and upper airway secretions. Anticholinergic drugs also reduce gut-motility, the normal **peristalsis**, intestinal movement during the digestive process.

Some pre-anesthesia drugs are given as a "cocktail," that is, combining two different injectable drugs into one syringe. These drugs are given IM approximately 30 minutes prior to general anesthesia. The time of delivery depends on the action of the drugs, how quickly they are absorbed and produce the desired effect, and the type of drugs used. Medications given prior to surgery can also function as post-surgical pain relief and assist in recovery from anesthesia by helping to keep the patient free from anxiety and pain.

The use and selection of pre-anesthesia drugs is determined by the veterinarian and depends on the assessment of the patient; underlying medical conditions, such as heart disease or kidney failure; age of the patient; and extent of surgical risk.

Surgical risk is a classification system developed by the American Society of Anesthesiologists, based on the physical status of the patient at the time of surgery (**Table 1-3**). Knowing the surgical risk, or *likely* surgical risk, benefits the veterinarian, the anesthetist, and the entire surgical team.

TABLE 1-3 Surgical Risk Classification.

DEFINITION	EXAMPLE
I A healthy patient presented for an elective surgery	Spay or neuter
II A patient with slight to moderate systemic changes	Dehydration, obesity
III A patient with major systemic changes that limit activity but are not incapacitating	Heart disease, anemia Severe fracture
IV A patient with very severe systemic changes that could lead to death without medical or surgical intervention	Frequent arrhythmia, ruptured bladder, internal hemorrhage, collapsed or punctured lung
V A patient in a critical state who will likely die despite medical or surgical intervention	Severe trauma with shock

INHALATION ANESTHESIA

Inhalation anesthesia means that the anesthetic gas (agent), in combination with oxygen, is delivered by a **vaporizer** and through a breathing system directly into the patient's lungs.

In a small-animal practice, two types of inhaled anesthesia agents are used—isoflurane and sevoflurane. Each has slightly different properties. Both are liquid until combined with oxygen through the vaporizer, where they become a volatile gas. Each must have a differently calibrated vaporizer; they are *not* interchangeable or ever used together in any combination.

Isoflurane is the most commonly used agent; its effects on the cardiovascular and respiratory systems are well known, and it is considered to have a great margin of safety. In general, isoflurane produces good muscle relaxation, has a slight analgesia affect, does not normally produce cardiac arrhythmias, and decreases blood pressure; 99% of iIsoflurane is eliminated from the body through the respiratory system. Isoflurane is also known by various trade names: Aerrane, Isovet, and Isoflo.

Sevoflurane has the advantage of a speedier induction and shorter recovery time. Advantages of sevofluane also include greater control over anesthetic depth and a more rapid response; however, great care must be taken when monitoring anesthetic depth and vaporizer settings to prevent a patient from becoming too deep too quick or becoming too light at a critical moment. Sevofluane does not produce any more cardiac depression than isoflurane but there is marked respiratory depression. It is better tolerated by patients during induction because the odor is low and there is reduced irritability to the airway. Vaporizer settings have to be higher, which means that a greater volume is used, and the cost difference can be substantial for a low-volume surgical practice.

ANESTHESIA BY INJECTION

Anesthesia also can be delivered by injection. It frequently is used for the induction of patients that are difficult to approach or for a quick procedure that may not require inhalation anesthesia—for example, a feline neuter. Short-acting injectable anesthetics facilitate intubation and can prevent the excitement phase of a patient being masked down. Injections of anesthetic agents are delivered either IV or IM, depending on the agent used. Generally, these drugs are classified as barbiturates (short- or long-acting), as non-barbiturates, or as dissociative agents. Not all of them provide analgesia and often are delivered in combination with another agent.

The decision to use injectable anesthesia is made by the veterinarian. All injectable agents are dose-dependent, so an accurate weight for the patient is critical in determining the correct dose. When using an injectable anesthetic, patients must be monitored just as carefully as they would be during a procedure using inhalation anesthesia. Although some of these agents can be reversed, the veterinary technician should be prepared to intubate the patient and provide manual ventilation if respiratory support is needed.

THE ANESTHESIA MACHINE

Every **anesthesia machine** has the same basic components: an oxygen source, a **pressure gauge** on the oxygen tank that shows the amount of oxygen in the tank, a **flow meter** that controls the precise percentage of gas delivery, and a vaporizer that converts the liquid anesthetic agent to the gas vapor. The anesthesia machine also incorporates a **flush valve** that is used to bypass the gas delivery outlet and fill the breathing system with pure oxygen and a pressure release valve, also called a **pop-off valve**, to prevent excessive pressure in the breathing circuit. Attached to the anesthesia machine and breathing system is a canister containing granules of soda lime that absorbs exhaled carbon dioxide (**Figure 1-68**).

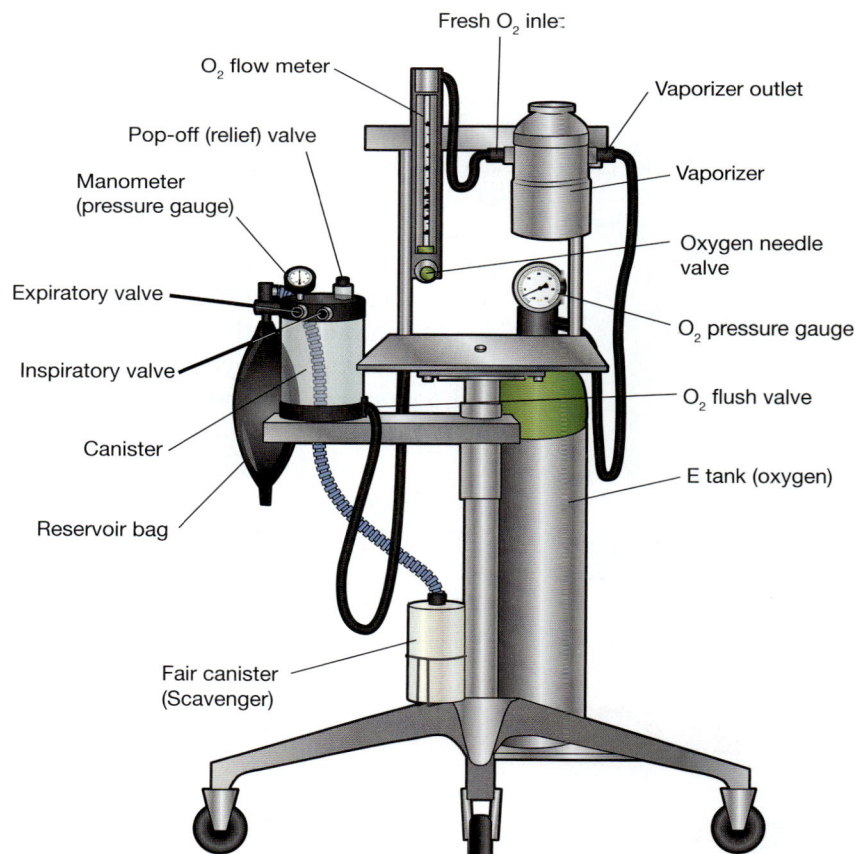

Fresh O_2 inlet

O_2 flow meter

Pop-off (relief) valve

Manometer (pressure gauge)

Expiratory valve

Inspiratory valve

Canister

Reservoir bag

Fair canister (Scavenger)

Vaporizer outlet

Vaporizer

Oxygen needle valve

O_2 pressure gauge

O_2 flush valve

E tank (oxygen)

▲ **FIGURE 1-68** A typical anesthesia machine used for delivery of an inhalent anesthetic gas.

The **breathing system** and an **endotracheal tube** (a tube placed within the trachea of the patient) connect the patient to the anesthesia machine for the delivery of oxygen, a controlled amount of anesthetic gas, and to provide a method of assisting ventilation in the anesthetized patient.

There are two types of breathing systems: the re-breathing system (circuit or circle system) and the non-re-breathing system. Re-breathing systems are most commonly used in veterinary anesthesia. They allow the patient to re-breathe the anesthetic gas and oxygen after it passes through a clear canister containing soda lime granules. The soda lime granules absorb exhaled carbon dioxide. The circle system incorporates two hoses, one for inhaled gas and oxygen and the other for exhaled gas and oxygen. The ends of each hose connect to the anesthesia machine ports, and the other ends, the "patient ends," are brought together with a Y connector. The Y connector is the link to the patient via the endotracheal tube. A re-breathing bag attaches to the anesthesia machine and provides the method for assisted ventilation.

SODA LIME CANISTERS

Fresh soda lime granules are white and contain a dye that changes the color of the saturated granules to light purple or blue. When two-thirds of the granules show this color change, the granules should be replaced, as they have reached the maximum saturation or absorption point.

Granules should be checked at the end of each surgery to determine if replacement with fresh soda lime granules is required. If the anesthesia machine is not used, the colored granules will eventually fade back to white.

Non-re-breathing systems do not have a canister for carbon dioxide absorption. Removal of carbon dioxide depends on the flow rate of both oxygen and anesthetic gas. These systems are used primarily in very small or young patients and in birds and reptiles (**Figure 1-69**).

The endotracheal (ET) tube is placed to ensure an open airway and delivers the anesthetic gas and oxygen directly into the patient's lungs. The ET tube is also necessary for assisted ventilation. ET tubes are either cuffed or non-cuffed. Cuffed tubes have an inflatable balloon that provides a seal within the airway and prevents aspiration. ET tubes are available in different sizes, both in length and diameter (**Figure 1-70**).

Every aspect of the anesthesia delivery system should be checked daily to ensure that it is ready for an emergency and *always* re-checked prior to use.

Use a checklist so nothing is overlooked. Never attempt to use an anesthetic machine that isn't functioning perfectly.

Checklist for Anesthesia Machine Function and Delivery System

1. Ensure that the oxygen supply is sufficient (open pressure valve).
2. Fill vaporizer with liquid anesthetic agent.
3. Check soda lime canister and change granules if necessary.
4. Attach breathing circuit and re-breathing bag.
5. Check for leaks:
   a. close pop-off valve
   b. occlude patient end, Y connector, of breathing system (circuit) with thumb
   c. fill circuit with oxygen to a pressure reading of 20 ml/minutes

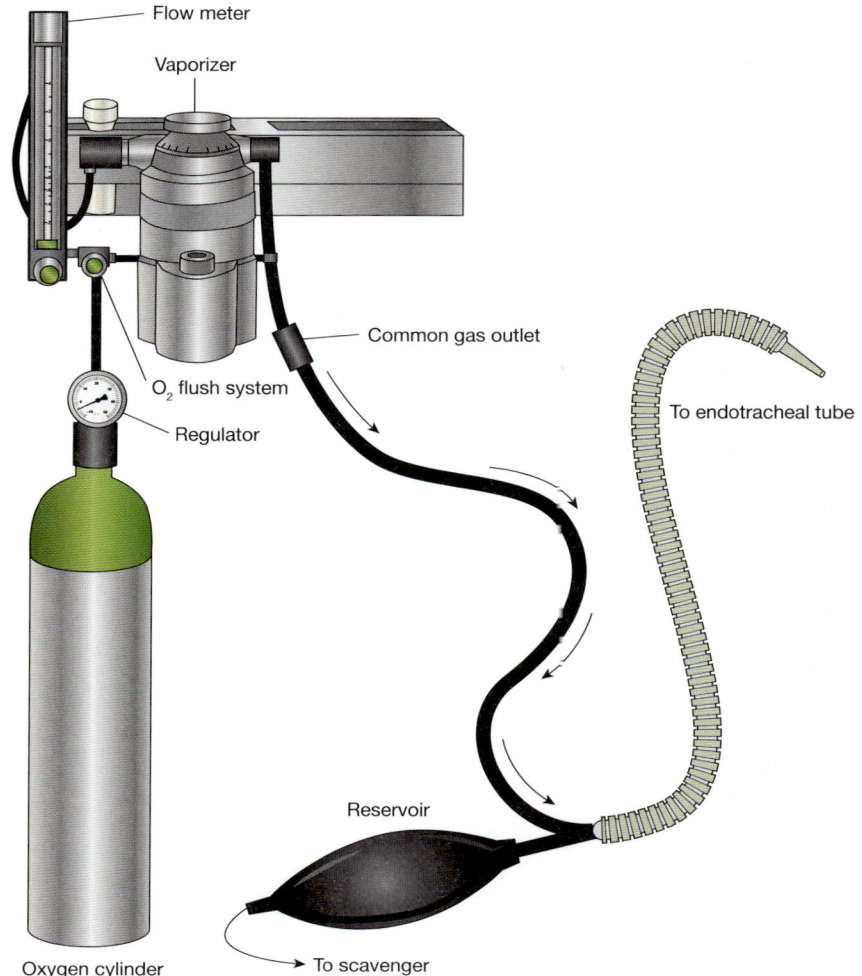

▲ **FIGURE 1-69** A non-re-breathing system.

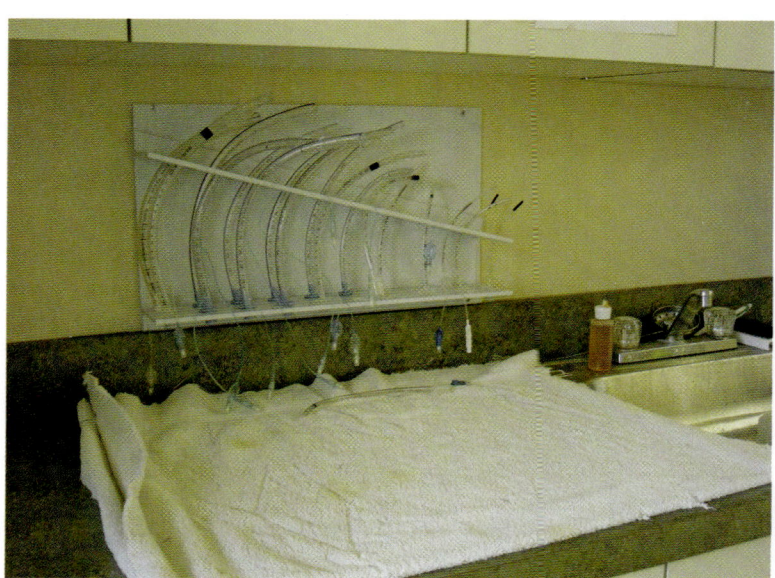

▲ **FIGURE 1-70** It is important to choose the correct size for the ET tube and discern between cuffed and non-cuffed ET tubes.

d. slowly increase oxygen flow to 100 ml/minutes

e. observe to see if pressure gauge increases (acceptable)

f. if pressure drops, increase flow rate of oxygen until the pressure gauge remains stable

g. leaks that exceed 200 ml/minutes are unacceptable and the machine requires professional maintenance. Do not use.

h. Open the pop-off valve while still occluding the breathing system. The pressure should drop to zero.

DEPTH OF ANESTHESIA

The depth of anesthesia must be monitored carefully during all stages of the procedure. Depth, or the plane of anesthesia, should never be assumed to be an "either/or" state. Many factors influence anesthesia depth: the absorption rate of pre-anesthetic drugs, the physiological status of the patient, and stimulation from the procedure. Anesthesia is not a constant state but, rather, one that fluctuates and is controlled by the anesthetist in anticipation and response to the patient and the procedure during the entire period of anesthesia.

Planes of anesthesia may be described as *light, surgical,* or *deep.* Anesthetic depth in the dog and cat can be assessed by evaluating muscle tone, eye reflexes, and jaw tone. Although these are extremely valuable, anesthetic depth in a patient must be assessed by all the physiological parameters including heart rate and respiration.

During light anesthesia there may be some spontaneous muscle or reflex movement, (withdrawal if toe is pinched) and the jaw tone is tight. The patient's eyes are central and the pupils are medium, neither constricted nor dilated. The **palpebral reflex** (blink reflex) is present, meaning that a light touch to the medial corner of the eyelids will cause the patient to blink.

During a surgical plane of anesthesia, spontaneous and reflex movement is absent. The jaw muscles are relaxed. The palpebral reflex may be only slight or non-existent. The eye position will move ventromedially (down and inward) and the pupil size is reduced.

During a deep plane of anesthesia, there will be no spontaneous or reflex movements, and the jaw will be slack. There is no palpebral reflex and the eye position has returned to the central position within the orbit. The pupils are dilated. *This depth is critical and, if not corrected immediately, could lead to the patient's death.* Immediate steps to lighten the plane of anesthesia are essential and include reducing the flow rate of anesthetic gas and providing manual ventilation until a surgical, adequate plane of anesthesia is restored. *It is imperative that all parameters of the patient's status are carefully and consistently monitored, and it is vital to anticipate and prevent a crisis—cardiac arrest and death from anesthesia depth.*

✳ SIGNS OF IMPENDING CARDIAC ARREST

- Cyanosis
- Changes in the respiratory rate and pattern (heavy abdominal breathing effort)
- Changes in the heart rate
- Pale mucous membranes
- Slow CRT
- Body temperature falls below 30° C (86° F)

THE ANESTHESIA RECORD FORM

Patient vital signs are recorded every 5 minutes with the use of a pre-printed **Anesthesia Record** form (**Figure 1-71**). This is a legal document, and though these forms may vary slightly from practice to practice, all provide for charting and recording the patient's vital signs. The anesthesia record provides an area for recording pre-medications, induction agents, fluids, and other drugs administered during surgery. The form, when used correctly, also produces a graph that charts the flow rates of anesthetic gas and oxygen. Although the information is recorded in 5-minute increments, it must never be forgotten that a serious incident could occur in a matter of seconds. Vigilance and the ability to respond immediately is required *every* minute. Doing nothing between the 5-minute intervals could be disastrous.

Anesthesia Record

Date:	**Procedure:**
Animal's Name:	**Species:**
Client's Name:	**Surgeon:**
Anesthetist:	**Assistant:**

Wt:	**Temp:**	**HR:**	**RR:**	**MM:**	**CRT:**

Physical Status: 1 2 3 4 5	**ET Tube:** __________mm	**Circle System**________ **Non-rebreathing System**_______

Pre-Anesthetic Drugs				**Anesthetic Induction Drugs**			
Drug	**Dose**	**Route**	**Time**	**Drug**	**Dose**	**Route**	**Time**

Keys: (I) Isoflorane (S) Sevoflorane (O) Resp (●) HR (*) SPO_2 (C) CO_2

Time:							
Gas: 1							
2							
3							
4							
5							
6							
200							
180							
160							
140							
130							
120							
110							
100							
90							
80							
60							
40							
20							
16							
14							
12							
10							
8							
6							
4							
2							
0							

▲ **FIGURE 1-71** Anesthesia Record Form.

All aspects of the patient's status must be constantly assessed, interpreted, and recorded. Primarily, the cardiovascular and respiratory systems are excellent indicators, but only if changes are recognized and acted upon immediately.

Cardiovascular function is monitored to ensure that cardiac output is maintained. It is assessed by the heart rate, heart sounds, mucous membrane color, pulse, and CRT. Blood must be circulating and carrying oxygen to all organs and body tissues. Each of the above and in combination help determine cardiac output.

Respiratory function is evaluated by the respiratory rate and breathing patterns. Without the use of specialized equipment, respiratory rate and patterns are observed by careful monitoring of the re-breathing bag. Movement of the re-breathing bag reflects not only the number of breaths per minute but also the character of those breaths: deep, shallow, rapid, and other abnormalities. Re-breathing bags should be matched to patient size.

Though it is not the focus of the surgeon to supervise the anesthesia, it is important to immediately advise him or her of any concerns. Changes in vital signs may result from manipulation of an organ or a particularly painful part of the surgery, but they also may predict an impending problem. Communication between the surgeon and anesthetist is important. *Never* "guess" that a problem will rectify itself.

It is strongly recommended that all students accompany and observe an experienced anesthetist through several procedures. When the use of the anesthesia record form and monitoring equipment is understood, the student may begin by observing and entering the data as called out by the anesthetist. Complete knowledge of the functions of the anesthesia machine also must be demonstrated under direct supervision. When both anesthetist and student are confident, the roles may be switched; the student should then, in the company of the experienced anesthetist, induce, maintain, and recover several patients prior to "flying solo."

Note: In most states only a credentialed veterinary technician may deliver anesthesia. New graduates of AVMA-approved programs will have had supervised anesthesia experience, and may already be certified; however, following the protocol above is still strongly recommended when starting in any new practice.

PROGRESSION OF ANESTHESIA

The first phase of anesthesia is **induction**. The induction phase is the beginning of transitioning from the effects of pre-anesthesia drugs to a loss of consciousness. Ideally, the transition to general anesthesia should be calm and without incident. The objective of the induction phase is to produce a loss of consciousness so endotracheal intubation can be performed. Without the use of anesthetic induction drugs, the patient is normally induced with the use of a face mask. This type of induction can cause an excitement phase and the patient may resist the face mask. Often, patients are induced with short-acting injectible anesthetic agents to provide for a smoother transition.

Anesthesia **maintenance** is the period following the induction phase when the patient is intubated and connected to the anesthesia machine. The maintenance phase continues throughout the entire period of anesthesia until recovery. Maintenance also means managing the level of anesthesia, carefully monitoring the patient, accurate recording of vital signs, and tracking, with the use of the Anesthesia Record, the physiological status of the patient with special regard to cardiac and respiratory functions. It is also important to monitor the performance of the anesthesia machine—the level of oxygen remaining in the tank, the level of the liquid anesthetic agent, and the patency of all tubes and connections.

The **recovery** phase of anesthesia begins when the procedure is complete and the patient has been removed from the anesthesia machine. Recovery can also be a critical time for the patient, and monitoring must continue until the patient is able to stand and there is no evidence of the effect of anesthesia or other drugs used. When signs of normal function begin to appear, the ET tube may be removed. The patient may blink and attempt to lick, but the ET tube should be removed only when the patient is able to swallow. In preparation for this, the ET tube should be untied and the cuff deflated. The recovery phase may include some vocalization and muscle movement, "paddling" of the legs, and premature attempts to rise.

INDUCTION of Anesthesia ("masking down")

TECHNICAL ACTION	RATIONALE/AMPLIFICATION
1. Position pre-checked anesthesia machine so that it is immediately available.	
2. Select appropriate-size face mask for patient, and attach to Y connector of anesthesia machine.	2a. To achieve maximum effect, the face mask should fit over the entire mouth and nose to prevent gas exposure to personnel. It should not be so small that it puts any pressure on the eyes.
3. With the assistance of a restrainer, place the patient in sternal recumbency on the table where induction and surgical preparation of the patient will occur.	3a. The patient should be relaxed but responsive. Be sure to give the pre-anesthesia drugs time to take effect.
4. Evaluate effectiveness of pre-anesthetic drugs.	
5. Place eye lubrication into each eye.	5a. The patient's eyes will remain open but non-blinking in a surgical plane of anesthesia. The lubricant is to prevent the corneas from becoming dry.
6. Turn on oxygen and allow to flow for a few moments.	6a. The oxygen always should be turned on prior to the vaporizer. Allowing a few moments of pure oxygen to flow through the circuit will remove any lingering odors and moisture that may have been retained within the breathing circuit.
	6b. During the induction phase, the flow rate of oxygen is set at 3L.
	6c. The induction flow rate, using isoflurane, is 1.5% to 2%. Flow rates are totally dependent on the patient status and should never be assumed as routine for all patients. Observing the patient and level of anesthetic depth is the most important indicator.
7. Gently position face mask over nares and mouth of patient and allow several breaths of pure oxygen.	7a. 8a. Allowing the patient to have a few breaths of pure—odorless—oxygen decreases the resistance to breathe when the odor of the anesthetic gas is introduced.
8. If the patient is calm and not resisting the face mask, introduce the inhalant anesthesia.	

(Continues)

TECHNICAL ACTION	RATIONALE/AMPLIFICATION
9. When the patient has reached a suitable plane of unconsciousness, have the restrainer position the patient for ET tube placement.	9a. 10a. 11a. Refer to Intubation Procedures in Surgical Prep.
10. Remove face mask and intubate.	
11. When intubation is confirmed and the ET tube is tied in, roll the patient onto lateral recumbency.	
12. Connect the ET tube to the anesthesia machine and begin monitoring vital signs during surgical prep: shaving and scrubbing the surgical field.	12a. Place the end of the ET tube in the receiving end of the Y connector on the breathing circuit. The oxygen is already flowing through the system and now directly into the patient's lung.
	12b. Take and record patient's heart and respiratory rate prior to introducing the anesthetic agent.
13. When surgical prep is complete, turn off anesthesia machine, vaporizer first, then oxygen, and smoothly transfer patient to the OR.	13a. If only one anesthesia machine is available, have an assistant wheel the anesthesia machine into the OR, putting it in position for immediate use.
14. Reconnect patient to the anesthesia machine in the OR, turn on oxygen and appropriate vaporizer setting, and continue to monitor vital signs.	14a. 15a. Review the guideline for determining the surgical plane of anesthesia. During maintenance, the gas flow rate should be adjusted to meet patient needs. Maintenance levels of isoflurane vary from 2.5% to 3%. Maintenance levels of sevofluarne can vary between 2.5% and 4.0%. There may be a need to increase the percentage for a short time if the patient is too light for the procedure. Always adjust the flow to meet the needs of the patient. Carefully monitor anesthesia depth when making any change to the flow rate, and record it on the Anesthesia Record Form.
15. Re-evaluate patient status and adjust flow rates as necessary to achieve a surgical plane of anesthesia.	
16. Do not, under any circumstances, leave the patient, not even for a moment.	16a. The anesthetist's primary function and only focus of attention is on the patient: to control the anesthesia, constantly monitor the patient, and monitor the equipment in use. An immediate response is not possible if the anesthetist has stepped aside to retrieve a stethoscope, a pen, or anything else. Always call for an assistant if something is required.

General anesthesia depresses respiration, and patients should periodically be given PPV to assist normal breathing and to maintain a stable flow of anesthetic gas and oxygen. PPV should be given at least once every 2 to 3 minutes throughout the period of anesthesia. The recommended respiratory rate for PPV is 8 to 14 breaths/minutes for both dogs and cats.

If the rate and depth of breathing is inadequate, the patient *must* receive positive-pressure ventilation (PPV). This is performed by closing the pop-off valve, and breathing for the patient by compressing the re-breathing bag. Compressions should match a normal breathing pattern. *Always reopen the pop-off valve after administering PPV.* Leaving a pop-off valve closed will result in the death of the patient.

Assisted ventilation (PPV) is given for two or three breaths in a regular sequence: Observe unassisted ventilation rate, close pop-off, give two or three breaths, *open pop-off valve,* and observe for unassisted ventilation. If the patient is not responding, repeat procedure. PPV should be used without hesitation whenever the patient is not breathing on its own.

RECOVERING A PATIENT FROM ANESTHESIA

Being aware of the progression of the surgery and communicating with the surgeon are two important aspects of anesthesia. When the procedure is nearly complete and the surgeon has begun suturing, he or she may call out, "closing." This is an indicator that the patient may be slowly brought to a lighter plane of anesthesia. It does not mean "finished," and the anesthetist is still responsible for maintaining the patient at the appropriate level. The percentage of gas may be reduced, usually by 0.5% increments but always with the needs of the patient first.

PROCEDURE for Recovering the Patient

TECHNICAL ACTION	RATIONALE/AMPLIFICATION
1. When the procedure is complete, begin the recovery by slowly reducing the anesthetic gas flow in 0.5% increments until the flow rate has be reduced to 0%.	1a. This allows for a smoother transition to recovery.
2. Allow the patient to breathe pure oxygen for a few moments.	2a. Breathing in oxygen perfuses tissues, stimulates normal respiration, and during exhalation removes or "blows off" residual anesthetic gas and carbon dioxide.
3. An assistant may untie the restraints and roll the patient onto lateral recumbency.	3a. The anesthetist must still remain focused on the patient and should not be moving around to untie the restraints.
	3b. The patient is placed in lateral recumbency because it is a normal resting position and easier to breathe.
	3c. Massage the limbs, especially at the point of tie-in, to assure and assist normal circulation.
4. Disconnect the patient from the Y connector on the breathing circuit and turn off oxygen. Do not extubate.	4a. The ET tube is left in place in case the patient requires PPV during recovery, and maintains the seal to the airway if the patient regurgitates or vomits and will prevent aspiration.
5. If the patient is breathing normally, carry the patient from the OR to the recovery area.	5a. If it is a large dog, request assistance.
	5b. Patients usually are placed on a padded area of the floor. Ideally, this could be a thick vinyl covered mat that can be washed and disinfected easily.
6. If the patient has been receiving fluids during the procedure, have an assistant bring the fluid stand and bag to the recovery area, being careful not to dislodge the catheter.	6a. If the patient has been receiving fluids during the procedure, it may continue during recovery.

(Continues)

TECHNICAL ACTION	RATIONALE/AMPLIFICATION
	6b. When the fluids are disconnected, be sure to flush the catheter and replace the cap. Leaving the catheter in place maintains venous access for medications and additional fluids if prescribed.
7. Lay the patient in lateral recumbency on a padded recovery area on the floor or in a padded cage.	**7a.** Lateral recumbency is a more normal position for sleep, and respiration is less restricted.
8. Deflate intubation tube cuff and untie gauze from around head and muzzle; leave ET tube in place.	**8a.** The cuff should be deflated in readiness for extubation. Never attempt to remove an ET without first deflating the cuff. Loosening the ties from the muzzle and head also prepares for extubation, but is still present if needed.
	8b. When the ties are loosened, examine the oral cavity and tongue for signs of abrasion if the tie-in has been inappropriately tight and pushed the lips against the teeth.
	8c. Check the CRT for color and perfusion. If the inner lips and tongue are very dry, a gauze sponge dampened with water may be used for moisturizing.
	8d. The ET tube should remain in place in case the patient requires PPV during recovery. Do not let a patient become apneic, but re-inflate the cuff and provide PPV until normal respiration resumes.
	8e. Leaving the ET tube in also prevents aspiration if the recovery patient regurgitates or vomits.
9. Keep the patient warm; cover with a blanket.	**9a.** Shivering occurs during recovery because the patient has lost body heat through the open incision. Do not just accept this as a normal part of recovery, but keep the patient warm with blankets, a towel-covered heating pad, or even surrounding the patient with disposable gloves filled with warm water.
10. Continue monitoring vital signs during recovery, and obtain rectal body temperature.	**10a.** Monitor until the rectal temperature returns to normal. Do not overheat the patient.
11. Monitor fluid bag for amount delivered and disconnect if total volume has been received.	**11a.** Refer to 6b.
12. Stimulate patient by speaking softly, using its name, and rubbing the sides of the body.	**12a.** Although the area of recovery should be kept quiet, stimulation of the patient is important. A patient may also be turned over onto the other side.
	12b. Some patients may exhibit paddling, rapid movement of the forelegs, and become very vocal. Comfort them, reassure them, keep a soft gentle voice.
13. Extubate when patient begins to swallow.	**13a.** Signs of an imminent recovery also include blinking and attempts to lick. Extubate only when the patient has swallowed. Rubbing the throat sometimes encourages swallowing.

(Continues)

TECHNICAL ACTION	RATIONALE/AMPLIFICATION
	13b. A patient must be able to swallow to avoid aspiration.
	13c. In brachaeocephalic patients (Pugs, Boxers, Persian cats) leave the ET tub in place until the last possible moment, that is, when the patient is fully alert and may be attempting to cough out the ET tube. Bracheocephalic breeds have longer soft palates, which could, until recovery is complete, block the airway.
14. Allow a few moments for the patient to re-orientate and attempt to sit up.	**14a.** Recovery, like induction, is a step-by-step process. The patient must never be rushed and allowed to readjust.
15. Return to clean prepared cage or kennel.	**15a.** The cage should be cleaned and disinfected, even if it is the same cage occupied prior to surgery. This removes as much bacteria as possible from the environment to keep the incision clean. Provide padding for comfort and warmth.
	15b. Often neglected is the removal of pre-surgical status cards such as NPO (nothing by mouth). If this is left in place, kennel staff will not offer food or water and the patient will be deprived.
	15c. When the patient is fully alert, a small amount of water should be offered. Observe the patient for regurgitation. Food and free access water usually can be provided in the late afternoon following full recovery.
16. Check incision site for signs of bleeding and abnormal swelling.	**16a.** Any abnormality or area of concern should be directed to the attention of the surgeon.

Surgical Preparation

© Liliya Kulianionak/Shutterstock.com

OBJECTIVES

Upon completion of this unit, the reader should be able to:

▶ Understand the admitting procedure for a surgical patient

▶ Prepare patient for anesthesia induction

▶ Perform endotracheal intubation in the dog and cat

▶ Perform surgical prep using the aseptic technique

▶ Position patient in the OR for surgery

KEY TERMS

aseptic technique

laryngoscope

nosocomial infection

NPO (nil per os)

scrub

solution

sterile technique

stylet

surgical release form

SURGICAL PREP

Most surgical patients are brought into the hospital early in the morning on the day of the surgery, or they may be dropped-off by the client the night before the procedure. When the patient is being checked in, a **surgical release form** is presented and explained to the client. The surgical release form is a legal document and must be signed by the owner of the dog or cat (or a representative of the client) before the patient is admitted. This form should clearly state the nature of the procedure to be performed, contain an acknowledgment of anesthesia risk, and a statement of understanding; that is, the procedure and risks must be explained to the client and understood by the client.

The form may vary from practice to practice, but many also include an *estimate* of cost. The estimate of cost is not the same as the quote of an exact all-inclusive price, but allows for the cost of additional treatment if required for the benefit of the patient, such as IV fluids and pain-control medications. If pre-surgical blood work and/or radiographs are recommended, this also needs to be confirmed with the client. In many practices, the veterinary technician is responsible for admitting surgical patients, and he or she should be prepared to answer clients' questions and concerns about the procedure and the purpose of the surgical release form.

When the patient is being admitted in the morning, the veterinary technician should confirm that food has been withheld; inquire if the patient is presently receiving any prescribed medications, supplements, or homeopathic treatments; and ask about any additional problems that the client has noticed. When a patient is presented for emergency surgery, the veterinarian usually consults with the client, explains what is involved—the risks, surgical options, and an estimation of costs. Regardless of the situation, the surgical release form must be signed by the client prior to the patient being admitted. The signed form is attached to the patient record.

Once admitted, the patient is given a complete physical examination. If a CBC and Bio/Chem profile are required, blood should be drawn at the time of the physical exam. Pre-surgical radiographs and those taken during and after the completion of an orthopedic surgery are normally done while the patient is anesthetized.

The admitted patient should be placed in a cage or kennel and the appropriate cage card completed. The cage card identifies the signalment of the patient and client information, the surgeon, and the procedure to be performed. In addition, the card should have a sticker affixed to it—**NPO**, which means "nil per os" or "nothing by mouth"—to avoid inadvertent access to food and water. Many practices also place a paper collar on the patient. These measures are taken to ensure that the correct patient receives the correct treatment and surgical procedure and there is no mix-up if the patient is moved to another kennel. All information should be copied onto the whiteboard designated for surgeries.

INTUBATION

Endotracheal intubation ensures a patent airway and provides for the ability to manually ventilate a patient when required. For dogs and cats, a cuffed tube is used. Once the ET tube is correctly placed, the cuff is inflated with air to fit within the trachea and prevent aspiration. Selection of the correct sized ET tube is important. Always choose the largest diameter that will slide easily into the trachea. The length should not extend farther than the thoracic inlet, nor should it extend more than 2 inches from the mouth. Sizes of ET tubes are measured by the internal diameter and are marked as ID and with the size in millimeters. Sizes available for dogs and cats range from 2.5 to 14 mm ID.

A **laryngoscope** is used to help visualize the opening to the trachea. The laryngoscope is a hand-held device with a light source and has a variety of different-size blades shaped for the curve of the animal's tongue and throat. It is placed in the patient's mouth to hold down the tongue while guiding the ET tube into the trachea (**Figure 1-72**).

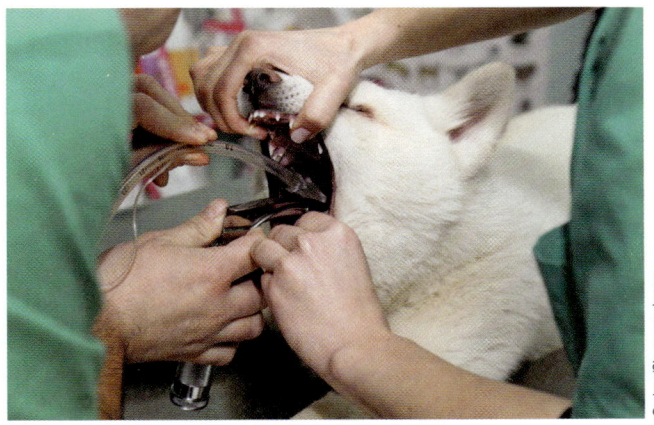

▲ **FIGURE 1-72** A laryngoscope is used to help visualize the trachea.

When using small ET tubes, a **stylet** is also helpful. A stylet is placed within the ET tube to give it more stability, and it is removed immediately once intubation has been achieved. Stylets should not extend beyond the end of the ET tube being placed to avoid trauma to the larynx. A metal probe should not be used because of the risk of puncturing the ET tube or the patient's throat. A stylet is particularly useful when intubating cats and small brachycephalic dogs, for example, the Pug and Lhasa Apso.

PURPOSE

- to provide a seal within the airway
- to prevent aspiration
- to provide PPV, positive pressure ventilation

COMPLICATIONS

- choosing the incorrect size ET tube
- not checking the cuff for patency
- not allowing enough time for pre-anesthetic agent to take effect
- poor restraint and positioning

EQUIPMENT

- Correct size cuffed ET tube
- 3 to 5 ml syringe
- Disposable gloves
- Gauze pads
- Length of rolled gauze
- Laryngoscope
- Restrainer

PROCEDURE for Intubation of the Dog

TECHNICAL ACTION	RATIONALE/AMPLIFICATION
1. Check the ET tube and cuff for patency.	**1a.** The ET tube should be smooth and clear of nicks and rough areas. The murphy eye, or opening of the ET tube, should not have any occlusions that could block respiration.

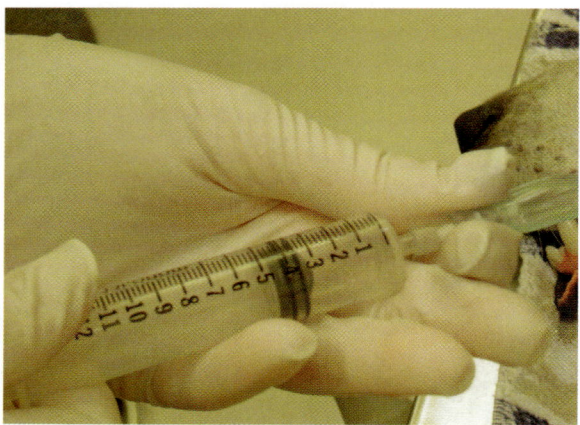

▲ **FIGURE 1-73** Air should be slowly injected from a syringe attached to the inflation port to visualize the amount of inflation necessary. The actual cuff is not visible in the trachea. Care must be taken not to over-inflate the cuff.

	1b. Cuffed ET tubes have a small-diameter length of tubing that is connected to the side of the ET tube. This has a cuff inflation indicator, a small "pillow" with a port to allow inflation of the cuff within the trachea (**Figure 1-73**). Cuff patency is tested by injecting air into the port and observing the cuff. The cuff should inflate and maintain the volume of air injected. To inject air into the port, attach the syringe to the cuff port and expel approximately 1–4 ml of air into the port (ml volume as marked on the syringe). The volume of air depends on the size of the ET tube. Do not over-inflate. When the cuff is determined to be patent, withdraw the air with the syringe to deflate the cuff.
	1c. Discard any ET tube that is non-functional to prevent it being used again. (Make a note of the size so it can be replaced).
2. Assess patient for the effect of pre-anesthetic sedation.	**2a.** The patient should be completely relaxed and free of anxiety. Do not administer additional pre-anesthetic medications, but use a face mask to administer inhalation anesthesia to achieve short-term but adequate level of sedation/unconsciousness for intubation.
	2b. Carefully evaluate jaw tone. The restrainer should be able to hold the mouth open without resistance.
	2c. The stimulation of placing the ET tube can rouse a patient, and the level of sedation must be adequate to prevent a bite injury to the restrainer or the person attempting to place the ET tube.
	2d. Do *not* call for additional restraint assistance; this will likely only increase the anxiety of the patient. If several people attempt to restrain the patient, the patient clearly is not ready for intubation. The patient should be comforted and kept as quiet as possible to allow the pre-anesthetic drugs to take effect.
3. Don gloves.	
4. The restrainer positions the dog in sternal recumbency and holds the mouth open.	**4a.** The patient's head should be held up, slightly extended, and in a straight line with the body.
5. Pull the tongue outward and down over the bottom incisor teeth.	**5a.** When grasping the tongue, place a fold of the gauze pad between the thumb and index finger for a better, non-slippery hold on the tongue.

(Continues)

TECHNICAL ACTION	RATIONALE/AMPLIFICATION
6. Visualize the opening to the trachea and insert ET tube.	**6a.** Turn the light on the laryngoscope and place the blade at the base of the tongue to push it down and out of the way for better visualization of the tracheal opening.
	6b. Position the ET tube in the curve of the blade toward the back of the tongue.
	6c. The trachea opens with inspiration. Wait until the patient takes a breath to insert the ET tube. Never try to force the ET tube into the trachea when it is closed.
	6d. Withdraw the laryngoscope (remember to turn off light source).
7. Confirm placement of ET tube within the trachea.	**7a.** Confirm placement of the ET tube by observing for a slight amount of condensation when the patient exhales. It is also possible to hear a breath being exhaled from the end of the tube. The restrainer then releases the mouth, allowing it to close.
8. Use the length of rolled gauze to tie the ET tube into place.	**8a.** When correct placement is confirmed, tie the length of gauze near the port end of the ET tube with a single knot, bring the ends up behind the canine teeth, cross them over, and tie across the muzzle. Then tie the remaining ends behind the head in a double bow.
	8b. Tying the ET tube keeps it in place so it can't be inadvertently dislodged when the patient is moved. Take care not to draw the gauze too tightly across the upper lips; it should be just snug enough to hold it in place.
9. Roll patient into lateral recumbency.	**9a.** It is easier for the patient to breath in lateral recumbency than to remain in sternal recumbency.
10. Inflate cuff.	**10a.** Inflate the cuff of the ET tube with air from the syringe.
11. Connect patient to anesthesia machine.	**11a.** When the patient is connected to the anesthesia machine, the re-breathing bag will inflate and deflate as the patient breathes.
	11b. If the person who is placing the ET tube is not the person who will be delivering the anesthesia, the anesthetist should be standing by and ready to begin the initial phase of inhalation anesthesia.

ENDOTRACHEAL Intubation of the Cat

Intubation of the cat is the same procedure as that for the dog with the exception of the following:

Gather the additional equipment needed:

- Stylet
- Lidocaine gel or liquid
- 1 cc syringe without needle
- Cotton-tipped swabs

TECHNICAL ACTION	RATIONALE/AMPLIFICATION
1. Follow steps 1–4.	
2. Just prior to step 5 and before attempting to insert the ET tube, apply a few drops (0.1 ml) of 2% lidocaine to the opening of the trachea.	2a. Cats have a very sensitive tracheal reflex. Lidocaine is a local anesthetic that numbs the area so it is not as sensitive.
3. Insert stylet into the ET tube.	3a. The stylet helps guide the ET tube and makes it less flexible. Once intubation is achieved, immediately remove the stylet.
4. Proceed with steps 5–10, but tying the ET tube only to the lower jaw, just behind the canines.	4a. The lower jaw is used to tie-in the ET tube because of the shape of the face and shorter muzzle of the cat.

ASEPTIC PREPARATION OF THE PATIENT

Surgeries are referred to simply as being "clean or dirty." An example of a "clean" surgery is a straightforward spay or any procedure that opens the abdominal cavity, whereas a surgery referred to as "dirty" may involve a patient with infected wounds or abscesses. Most hospitals determine the order of the surgeries beginning with "clean" procedures and finishing with "dirty" procedures to avoid possible contamination of the surgical suite and to reduce the likelihood of a **nosocomial infection**. A nosocomial infection is one that is acquired in the hospital through exposure to bacteria in a contaminated environment. Sources of contamination include surgical staff, the operating room, surgical instruments, and equipment, and even the patient when **aseptic technique** is inconsistent or lacking.

Aseptic technique refers to all precautions taken to prevent contamination, especially in preparing the incision site. It is different from, but no less important than, **sterile technique**, which is the absence of all microbes. During a surgical procedure only sterile items may touch tissue. This includes not only the instrument pack but also the gloves and gown of the surgeon, the surgical drapes, and suture material—in short, any item that comes in contact with the patient's tissues.

Items in the OR should be separated into sterile and non-sterile areas, and all staff members involved need to know the difference to avoid contamination. This is especially important when transferring the patient from the prep area to the OR table. If there is ever any question that an item is not sterile, it should be considered "dirty" (contaminated) and replaced. Aseptic technique is used when scrubbing and preparing the surgical site, and for this reason, all patient prep is done in an area separate from the operating room. Surgical preparation (clipping and scrubbing) begins when the patient has been anesthetized and intubated, the OR has been set up, and procedures are on schedule.

PURPOSE
- To clip and scrub the surgical site
- To use aseptic technique and prevent contamination

COMPLICATIONS

- Preparing the wrong patient
- Not adhering to aseptic technique
- Clipper burn
- Amputation of nipple with clippers
- Contamination of site during transfer to the OR

EQUIPMENT

- Electric clippers
- Number 10 clipper blade
- Number 40 clipper blade
- Hand-held or central vacuum system
- Water-soluble gel
- Gauze pads soaked in surgical scrub
- Gauze pads soaked in surgical solution
- Gauze pads soaked in isopropol alcohol
- Spray bottle containing isopropol alcohol
- Disposable gloves
- Plastic disposal bag for used gauze pads

ASEPTIC Preparation of the Surgical Patient

TECHNICAL ACTION	RATIONALE/AMPLIFICATION
1. Confirm patient; confirm procedure.	1a. Always confirm that it is the correct patient, even though this was likely done when anesthetized and intubated.
2. The surgical site is first clipped with a #10 blade to remove long hair. The coat is shaved in the direction of hair growth.	2a. The procedure determines the area that is clipped and prepped.
	2b. If the surgery is on a limb, the entire limb should be clipped and scrubbed. To facilitate this, the limb is hung up with tape stirrups and gauze.
	2c. For a feline castration, the testicles are not shaved, but the hairs are plucked from the scrotum.
	2d. For a canine castration, great care must be taken when shaving the scrotum. The skin is very sensitive and a clipper burn or abrasion will be irritating to the patient, causing the dog to lick the scrotum and incision site. Licking the incision site causes healing to be delayed.
	2e. For a feline declaw, the paws are not shaved. Each toe and the pads of the feet should be scrubbed, rinsed, and dipped in a basin of chlorhexadine solution.
	2f. Clip the hair with a #10 blade in the direction of hair growth. Doing this makes shaving the area easier and outlines the boundaries of the area to be clipped and prepped. As always, keep the edges straight and neat.

(Continues)

TECHNICAL ACTION	RATIONALE/AMPLIFICATION
3. Vacuum up loose hair.	**3a.** Using the vacuum frequently keeps the area free of clipped hairs and contamination of scrub and solutions.
4. Change the blade to # 40.	**4a.** The # 40 blade will shave the hair to the patient's skin. Check the temperature of the blade frequently by laying it on your own inner arm. If it feels too hot, it will also be too hot for the patient's skin. Either change the blade or cool it off using Cool-Lube® spray or by running the clipper in a basin of solution of dilute chlorhexadine.
	4b. All blades should be sharp, clean, and without missing teeth. Damaged, dirty blades become hot more quickly and are a source of clipper burn, nicks, and cuts. Be careful not to amputate a nipple, especially when using the # 40 blade. Nipples are present in both sexes.
5. Shave the surgical area in all four directions away from the incision site: cranial, caudal, and both sides laterally. When using the # 40 blade, the hair is shaved in the opposite direction of hair growth,	**5a.** With the # 40 blade, the hair is shaved against the hair growth to lift it up and make the area as free of hair as possible.
	5b. The margins of the shaved area should extend 2 to 4 inches beyond the incision site.
6. Vacuum all shaved hair from patient and surrounding area.	**6a.** Small clippings of hair are less visible. Use the vacuum over the skin surface of the patient, clothes of personnel, and the surrounding area.
7. In long-haired cats and dogs, apply a sterile water-soluble gel to the edges of the shaved area.	**7a.** A thin application of water soluble gel to the margins of the shaved area will keep long, untrimmed hair away from the surgical field.
	7b. If the patient has an open wound, fill the area with the gel. It will keep hair clippings out of the wound and can be rinsed away with solution or 0.9% saline when shaving has been completed.
8. Don gloves.	**8a.** Disposable gloves should be worn to reduce the possibility of a nosocomial infection. The gloves become contaminated when prepping an infected wound, and should be changed. Discard and use a new pair before continuing with the patient scrub.
9. Scrub the entire shaved area outward from the center in a spiral, never crossing back to the area just scrubbed (**Figure 1-74**) Continue until no dirt or debris appears on the gauze pads.	**9. 10a.** The process of scrubbing the patient involves two steps, repeated a minimum of three times each. **Scrub** is a soapy liquid, and the **solution** contains no sudzing agents.
10. When the used gauze pads appear clean, repeat the same procedure with either a chlorhexadine or betadine solution, and finish with isopropol alcohol.	**9. 10b.** The two most commonly used agents are chlorhexadine (Novalsan) and betadyne, which is iodine-based. Chlorhexadine is blue, and betadyne is rust-colored and will stain the hands and skin. Both are available as a scrub or a solution.

(Continues)

TECHNICAL ACTION	RATIONALE/AMPLIFICATION

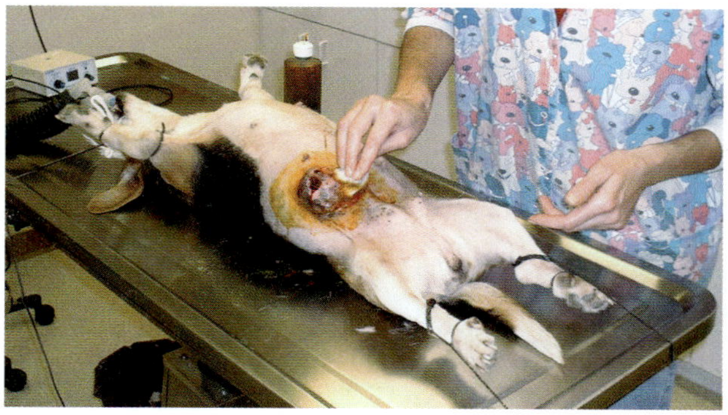

▲ **FIGURE 1-74** The aseptic technique should be adhered to in preparing the surgical site.

9. 10c. Isopropal alcohol is an antibacterial agent. It is used after the scrub has been completed or as one of the alternating agents between the scrub and the solution. The spray bottle application prevents touching the aseptically prepared surgical site. It should be applied again once the patient is in position on the OR table.

9. 10d. Never use isopropol alcohol in an open wound. It is very irritating to tissues.

11. Repeat a minimum of three times, alternating between scrub and solution or scrub and alcohol.

12. Using a spray bottle of solution or alcohol, wet the entire scrubbed area.

13. Transport the patient to the OR table and position for surgery.

13a. When transporting the patient to the OR, take care not to contaminate the surgical field. If this occurs, re-scrubbing is required.

13b. Large patients can be transported on a wheeled cart. An assistant should always be available to steady the patient. In addition, the anesthetist should be present to disconnect the patient from the anesthesia machine in the prep area and re-connect the patient in the OR.

13c. If fluids are connected and running, the fluid stand can either be transported with the patient or the fluid line disconnected and re-started once the patient is on the OR table.

13d. Care must be taken by all personnel not to contaminate the surgical field, dislodge the IV line and catheter, and be aware of the location of all sterile items placed in the OR.

14. Once positioned on the OR table, re-spray the incision site with solution or alcohol used in step 12.

POSITIONING AND SECURING THE PATIENT

The type of surgery determines positioning of the patient. During a fracture repair, the patient is placed in lateral recumbency, with the uninjured side down. For an abdominal procedure, the patient is placed on its back in dorsal recumbency. Canine neuters are also positioned in dorsal recumbency. Feline neuters are often performed in the prep area and the patient is presented in ventral recumbency with the caudal area elevated.

All patients, with the exception of the feline neuter, are secured to the surgical table with soft cotton rope. Every table has adjustable "tie-downs" to accommodate different-sized patients (**Figure 1-75**).

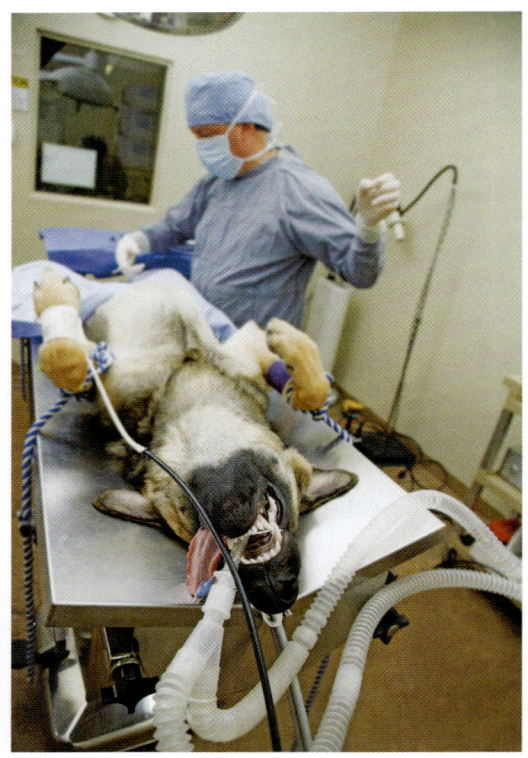

© CREATISTA/Shutterstock.com

◀ **FIGURE 1-75** The patient is correctly tied and positioned in a V-trough table.

All four limbs are secured, with the exception of an injured limb. This helps to keep the patient in a stable position and prevents the patient from movement during the procedure, which is especially important if the plane of anesthesia temporarily becomes too light. The patient should not be stretched so far as to force an unnatural position by splaying the rear legs unnaturally or the forelimbs so extended so that respiration is compromised.

Many tables also are designed to form a V trough, which helps support the patient in dorsal recumbency. Regardless of the type of table used, all patients should be supported with the use of foam padding, often cut into wedges and used depending on patient positioning.

All patients should be provided with an external heat source such as a circulating warm water bed. For very small patients, disposable gloves can be filled with warm water to surround the patient. Electric heating pads should not be used. Preventing hyperthermia is also very important.

Radiology

OBJECTIVES

Upon completion of this unit, the reader should be able to:

▶ Explain radiation safety
▶ State the reasons for radiographs
▶ Maintain the X-ray log
▶ Apply terminology used in radiology
▶ Describe and demonstrate the settings on an X-ray machine
▶ Explain how images are captured with film or by using digital radiology techniques
▶ Demonstrate correct positioning for V/D and lateral views
▶ Take diagnostic radiographs of the dog and the cat
▶ Correctly handle exposed film
▶ Imprint information onto a radiograph
▶ Re-load the film cassette
▶ Correctly file radiographs taken with film
▶ Develop radiographs using an automatic processor

KEY TERMS

artifact	developer	radiology	technique chart
automatic processor	direct radiography (DR)	radiology log	X-rays
computed radiography (CR)	dosimetry badge	radiolucent	X-ray film
contrast	film cassette	radiopaque	
darkroom	fixer	safelight	
	intensifying screen	scatter radiation	

© Liliya Kulianionak/Shutterstock.com

107

RADIOLOGY

Radiology is one of the most frequently used diagnostic procedures in veterinary medicine. The picture or film produced can determine the surgical approach, the type of fracture or blockage a patient may have, and may reveal unknown problems.

A radiograph is produced when **X-rays** penetrate the exposed area and are captured on a sheet of specialized X-ray film or by the use of digital imaging. The result is a permanent portrait of the area that has been radiographed and appears on the developed film or in the digital image much like any photographic negative.

OSHA (OCCUPATIONAL HEALTH AND SAFETY ACT, 1970)

Exposure to radiation (X-rays) is harmful. Radiation kills cells, causes reproductive failure, damages DNA, and can cause mental retardation and congenital malformation in an unborn fetus. In addition, radiation exposure can affect blood and lymph cells, thyroid function, and damage to the lens of the eye. Because X-rays cannot be detected by any of the senses, all staff members working in radiology must understand the importance of and adhere to radiation safety regulations.

No one under the age of 18 is allowed to work in radiology, and women who are pregnant, or think they might be, should advise the supervisor immediately. Members of the public (clients) are not allowed to remain in the area when a radiograph is being taken, and the only people who should remain in the room are trained staff restraining the patient and taking the films. Anyone in the room must wear PPE (personal protective equipment).

Radiation safety begins with understanding the dangers and always using protective equipment. A personal **dosimetry badge** is required for each staff member who works in radiology (**Figure 1-76**). Dosimetry badges absorb and retain an individual's exposure level. These should be clipped onto the front of a garment or the lead apron being used. Periodically, the badges are replaced

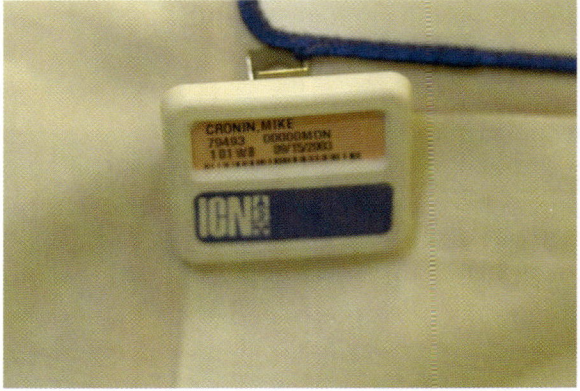

▲ **FIGURE 1-76** A dosimetry badge should be worn by every member of staff while taking radiographs.

and the exposed badges returned to a company that measures and reports radiation exposure for each individual. No one should use a badge assigned to another member of the staff and, equally important, is to always wear this personal monitor while working in radiology.

When the badges have been measured, the company reports back to the facility with the level of exposure for each individual. When not in use, dosimetry badges should be stored in an area away from the X-ray room and not exposed to sunlight.

All staff members participating in radiology should wear a protective lead apron, lead gloves, a thyroid shield, and lead glasses. The lead in this equipment shields the user from exposure to radiation with the exception of the primary beam. No part of an unprotected person should ever appear in a radiograph. Careless use of gloves during restraint is the usual culprit when a perfectly radiographed human finger or thumb bone appears on the developed film.

In addition, the equipment used should be radiographed periodically to check for cracks or damage in the lead shield. Damage can occur when the items are folded. Aprons should always be stored on hangers and the gloves returned to the glove rack. Thyroid shields and lead glasses should also be handled carefully and stored appropriately to prevent damage.

Scatter radiation is the number-one source of exposure. Scatter occurs when the main beam of the X-ray hits a solid surface (including the patient) and "scatters" in every direction (**Figure 1-77**). Methods of minimizing exposure to scatter radiation include:

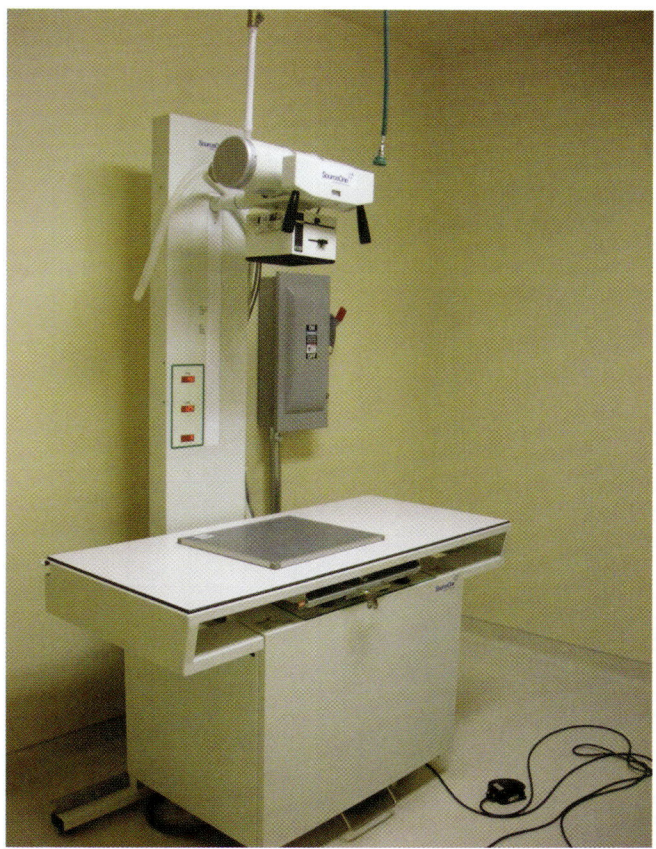

▲ **FIGURE 1-77** Scatter radiation occurs when X-rays strike a solid surface.

- If using film cassettes, avoid retakes; this can be caused in part by patient movement, inadequate restraint, and poor patient positioning
- Determine the correct settings for film exposure
- Accurately measure the area to be radiographed
- Use the smallest film cassettes; do not use a 9 × 12 cassette if a 5 × 7 is adequate for the area
- Use the collimator to reduce the size of the primary beam
- Always wear protective lead shielding
- Keep all body parts out of the primary beam
- Increase your distance from the primary beam
- Minimize individual exposure time through staff rotation
- Use extensions for restraint; lengths of gauze or soft cotton ropes

THE RADIOLOGY LOG

Maintenance of a **radiology log** (or X-ray log) is a legal requirement and must be completed when every X-ray is taken. Different states determine what information is required. This log includes client and patient information, the number assigned to the radiograph, details and views of the body part radiographed, and the machine settings. Depending on the form used, there also may be a space for comment on film quality and the veterinarian's diagnosis. The log should be kept in a binder in the radiology room (**Figure 1-78**).

X-RAY LOG													
Date	X-Ray #	Client Name	Patient Name	Species	Sex	Age	Weight	Veiw	CM	KVP	mAs	Body part	Initials
								-----	-----	-----			-----
								-----	-----	-----			-----
								-----	-----	-----			-----
								-----	-----	-----			-----
								-----	-----	-----			-----
								-----	-----	-----			-----
								-----	-----	-----			-----
								-----	-----	-----			-----
								-----	-----	-----			-----
								-----	-----	-----			-----

▲ **FIGURE 1-78** Radiology log.

FILING FILM RADIOGRAPHS

All radiographs taken with film must be filed in an area of the radiology room. Every radiograph taken for that patient is kept within one assigned file, usually an open-ended sleeve, and filed numerically. The outside of the sleeve should contain the same information that was recorded in the radiology log. X-rays, whether produced digitally or by film, are part of the medical history of the patient and should be treated with the same confidentiality.

Films or digitally captured images stored in a computer file should never be given to anyone, with the exception of a consulting veterinarian. Radiographs are the property of the practice, not the client. On some occasions, clients may ask to take films with them if they are moving or leaving the practice. This request should be politely denied and an explanation provided. Depending upon the practice, clients may obtain copies at their own expense, but not the original films, which are required by law to be retained by the practice. With the use of digital radiology, images are quickly and easily forwarded to a consulting veterinarian and, depending upon the protocols adopted by the practice, they also may be sent as an e-mail attachment to the client.

X-RAY FILM AND CASSETTES

X-ray film consists of three layers: a protective layer of an emulsion gelatin, an emulsion layer with silver halide crystals, and a polyester film base that coats both sides of the film. When exposed to radiation, the silver halide crystals become sensitive to the chemical change that occurs during developing. The latent image captured on the film is changed to a visible image that appears in black and white and shades of gray, described as **contrast**. Contrast is the difference between the densities of the body parts being radiographed. The least radiodense image is air. For example, a pocket of air in the intestinal tract will appear black. Bone and metal are the most radiodense and appear as white to bright white. Black and dark areas that appear on developed film are described as **radiolucent**. Areas that appear light or white are described as **radiopaque** (Figure 1-79).

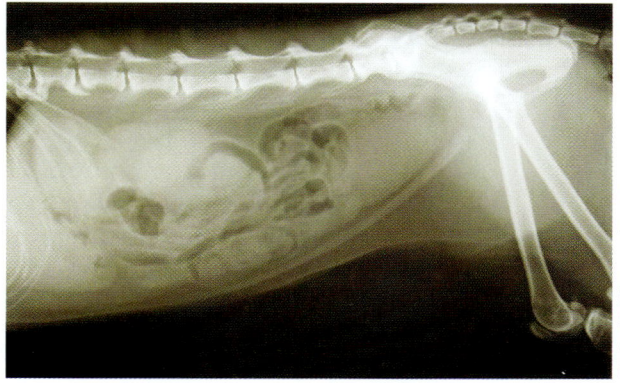

▲ **FIGURE 1-79** A lateral radiograph of a dog showing the different densities of body tissues.

Boxes of unexposed film should be stored in the darkroom in a film bin. Boxes of film should be stored either on end or flat on the side. Film is sensitive to pressure, so boxes should not be stacked one on top of another. Film should always be handled carefully and held by the edges to prevent damage and possible **artifacts** that appear on the developed X-ray. An artifact could be caused by a scratch or crease in the film, or dust, and is any item that appears on the developed film that is not a part of the patient.

Film casettes contain **intensifying screens**. The sheet of film is placed within the same-sized cassette. Cassettes are light-proof and have built in intensifying screens. Intensifying screens contain fluorescent crystals. When exposed to X-rays, the crystals emit light and record the X-rays that penetrate the patient and increase the photographic effect. The cassette holds the film in close contact with the two screens like a sandwich. Cassettes should not be dropped or handled roughly and should be examined regularly for damage. Cassettes and the screens should be cleaned regularly as dirt and dust can also contribute to artifacts appearing in a developed film (**Figure 1-80**).

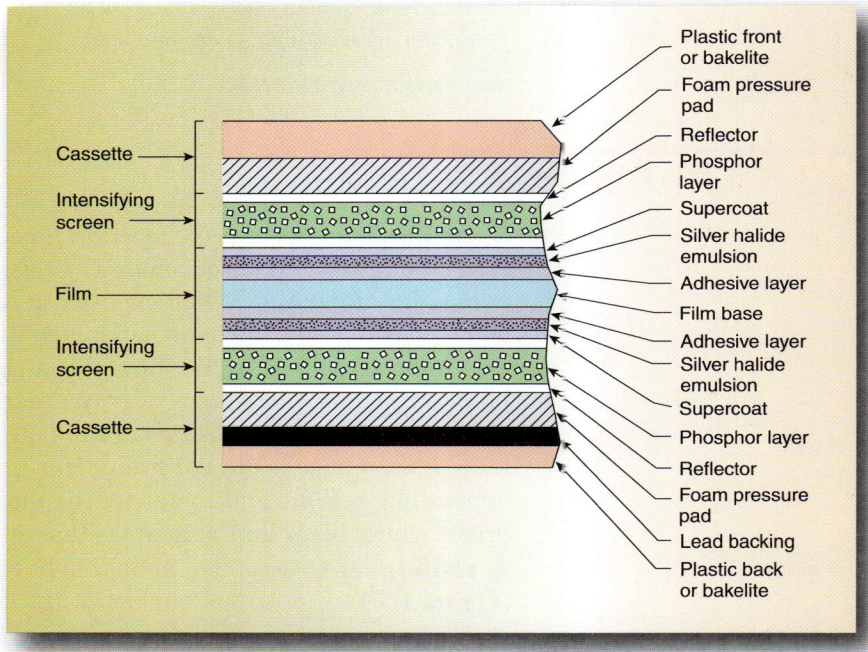

▲ **FIGURE 1-80** Cross-section of an X ray film cassette.

DIGITAL IMAGING

Digital imaging, or **computed radiography (CR)** technology, has replaced the use of film and film cassettes in many practices, and this technology is advancing rapidly. Computed radiography replaces X-ray film with an imaging plate built into a cassette. Photosensitive chemicals (phosphors) within the plate react when scanned with a laser to capture the latent image. After exposure, the cassette is placed in a reader that records and produces a digital image (the radiograph) on the computer.

One of the newer technologies is **direct radiography**, and it is similar to computed radiography but without the need for placing a cassette into a reader.

The image is captured directly on a built-in flat plate, and the image is transmitted immediately to the computer.

Digital radiography, both direct (DR) and computed radiography (CR), have many advantages:

- X-ray sensors are used instead of costly X-ray film
- Chemical processing is not required, eliminating developing time and chemical exposure to staff
- No darkroom is required
- Less radiation is generated
- Images are immediately available for viewing
- Images can be enhanced, modified, or deleted
- Radiographs can be sent to a network of computers
- Images are filed on computer discs for easy retrieval

GUIDELINES FOR TAKING RADIOGRAPHS

Review all radiation safety procedures and use safety equipment correctly. Understand the settings on the X-ray machine, and measure the patient accurately. In most instances, a minimum of two views are required: ventral/dorsal (VD) and lateral (Lat). For a VD view, the patient is on its back and the X-ray beam penetrates through the abdomen to the spine. For a lateral view, the patient is placed on its side, and always "bad side" down. In other words, a suspected fracture of the limb should be placed closest to the cassette. If the patient is difficult to restrain or there is pain, or likely to be pain, the patient should be anesthetized. A quick "mask down" with an inhalant anesthesia is necessary.

PURPOSE
- To obtain radiographs of diagnostic quality

COMPLICATIONS
- Inadequate restraint
- Incorrect positioning
- Incorrect exposure
- Incorrect measurement of patient
- No film in cassette
- Damaged cassette
- Incorrect software installed for type of computer imaging capture
- **R** or **L** lead markers omitted

EQUIPMENT
- Measuring rule in centimeters
- Technique Chart
- Correct size cassettes
- Correct sized panel to capture image
- **R** or **L** lead markers
- Assistant for restraint

BASIC Procedure for VD and Lateral Radiographs

TECHNICAL ACTION	RATIONALE/AMPLIFICATION
1. Measure the thickest part of the area to be radiographed (**Figure 1-81**).	1a. The thickest part of the area to be radiographed should be accurately measured in centimeters. Slide the fixed arm of the measuring stick underneath the patient at the thickest area to be radiographed. Lower the movable arm so it touches the body part but without any pressure on the area. Record the centimeter reading just below the movable arm.

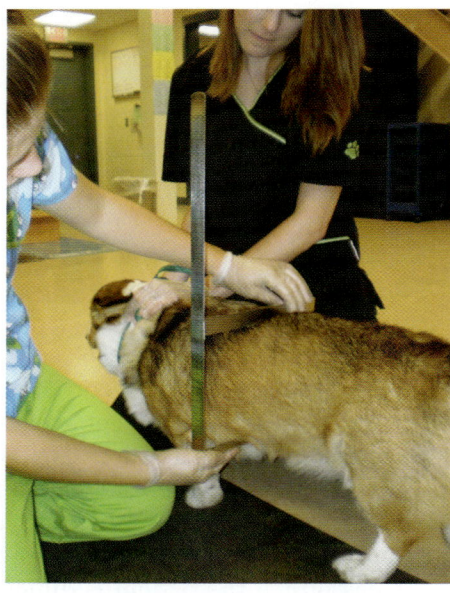

▲ **FIGURE 1-81** Correct measurement of the patient is necessary to adjust the settings on the X-ray machine.

TECHNICAL ACTION	RATIONALE/AMPLIFICATION
2. Consult Technique Chart for machine settings.	2a. A **technique chart** lists the settings for the X-ray machine. It is set up in columns based on the centimeter measurement, and the columns designate the kVp and mAs setting that should be used for that specific machine. These may have to be modified for use with digital radiographic techniques.
	2b. There are different technique charts depending on the placement of the film cassette: tabletop, directly underneath the patient, or if the cassette is placed in a grid that holds the cassette underneath the table. The grid system reduces scatter radiation and also increases contrast but requires a higher kVp setting. Grids are used most frequently when the area being radiographed is greater than 10 centimeters in thickness.
3. Adjust collimator to decrease range of exposure from the main beam.	3a. By reducing the size of the main beam, exposure to personnel is lessened and focused more on the area to be radiographed.
4. Don all safety equipment: lead apron, lead gloves, thyroid shield, and goggles.	

(Continues)

TECHNICAL ACTION	RATIONALE/AMPLIFICATION

5. Place film cassette or digital panel on the X-ray table, and place lead marker to determine right side from left side (**Figure 1-82**).

▲ **FIGURE 1-82** Lead markers are placed directly onto the film cassette or digital plate to distinguish right from left in the radiograph.

6. Position patient for V/D view.

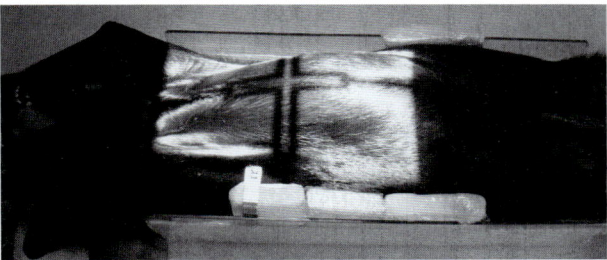

▲ **FIGURE 1-83** Correct positioning for V/D view.

7. Adjust location of film cassette or digital panel so it is directly under the target area to be radiographed.

8. Use the foot pedal to start the rotator.

5a. Unless the view is designated right or left, it may be difficult to distinguish once the radiograph is developed.

6a. For a V/D view, the patient is placed on its back with limbs extended. The forelegs are extended cranially and the hind limbs are extended caudally. The head should be placed between the forelegs and the tail between the hind limbs (**Figure 1-83**). Extending the limbs prevents overlay, that is, when a limb is superimposed over the top of organs and obscures the internal organs and spine. Patients should always be straight and in close contact with the cassette or digital plate and tabletop.

6b. Unless there is a left-limb injury, the patient is usually placed on its right side. An injured, fractured limb should always be placed down, directly onto the cassette.

7a. Always ensure that the cassette or digital capture plate is directly underneath the target area and correctly aligned with the area to be radiographed. If not positioned correctly, the image may reflect only part of the area that should have been radiographed.

8a. X-ray machines have a two-step exposure switch. Most machines operate with a foot pedal, which should be pulled as far away from the table as possible but still allow the patient to be restrained correctly. This is not only a safety measure but also allows for a last-second stop in the release of X-rays.

8b. Step one of the exposure switch starts the rotation of the anode. It can be heard as a light whir when the foot pedal is pressed lightly. This does not release X-rays. Stepping down completely on the foot pedal activates the high

(Continues)

TECHNICAL ACTION	RATIONALE/AMPLIFICATION
	voltage circuit, the kVp, X-ray production, and release of the X-rays. When the foot pedal is completely depressed and X-rays are released, the machine produces a short buzzing sound or another distinct sound that confirms that the X-rays have been released and exposure is complete.
9. Lean back and turn away as far as possible and look away.	9a. This is a personal safety measure. Turning your face helps protect the eyes. Never look straight in the X-ray field when exposure occurs.
10. Step down completely on the foot pedal to release X-rays.	10a. Stepping down completely releases the X-ray beams. Refer to 8a and 8b.
11. If using film, remove exposed cassette from under the patient, remove L or R marker, and set cassette aside, away from the table.	11a. The markers will be needed for the lateral view. By moving the exposed film safely away from the table, it not only avoids a mix-up in which cassette has been exposed, but also potential contamination if X-ray beams from a subsequent radiograph exposes it to scatter radiation.
12. Re-position the patient for a lateral view.	12a. Refer to 6b.
13. Follow steps 1-11 above.	

DEVELOPING FILM

Developing the exposed X-ray film is the final step in producing a diagnostic radiograph. If an exposed film is mishandled in the darkroom and the developing process is not understood, the films, even with the best radiographic technique, can easily be rendered useless for diagnostic purposes. Incorrect processing of the films creates the need for re-takes, causes frustration, and more handling of a patient that may already be stressed. It also costs time and money, and X-ray film is expensive.

In almost all veterinary practices, films are developed by an **automatic processor** rather than the more time-consuming method of hand tank developing that was used previously.

Films are developed with chemicals that change the latent image into a visual image and preserve the radiograph to create a permanent record. The chemicals in the **developer** solution change the silver halide crystals in the film to black metallic silver. The **fixer** solution removes all the silver halide crystals that have not been changed by the developer, leaving only black metallic silver and producing the image. The fixer also contains chemicals that preserve and harden the film.

Developers and fixers are available either in powder form or as a concentrated liquid, and both are made into solutions with water. When working with chemical solutions, care should be taken. Drips and splashes can not only ruin a film but can also cause eye and skin irritation.

Boxes of film and cassettes holding film are opened only in the **darkroom**. The darkroom is sealed to prevent any light from entering. Most darkrooms are

Courtesy of Air Techniques, Inc.

▲ **FIGURE 1-84** Automatic film processors have replaced the hand tank developing of films, saving time and reducing exposure to harmful developing chemicals.

located within a small area of the radiology room. Automatic film processors are compact and fit easily onto a countertop in the darkroom and take up far less space than the tanks for hand developing. (**Figure 1-84**)

All darkrooms have a **safelight**, a type of bulb that provides sufficient illumination but does not harm the film. Safe lights also have filters to block out specific light rays that destroy film.

When someone is using the darkroom, there is also a warning light that should be turned on to prevent people from opening the door accidentally. Normally, this light is outside of the entrance to the radiology room and should always be turned on to alert staff that the darkroom is in use. Many darkrooms have an additional inner latch to ensure that the door remains closed during film processing.

There are several models of automatic processors for developing film, but they all function in basically the same way and contain the same type of chemicals for developing and fixing. The film is automatically pulled through a series of rollers to the developer, the fixer into the water bath, and onto a dryer. The advantages are that films can be developed and read more quickly, and they eliminate the need for re-takes because of processing errors.

Users should become totally familiar with the manual provided for the specific make and model in the practice. The manual details all the parts, and recommendations for the type of film. It also recommends maintenance and cleaning schedules. The cleaning schedule is an important aspect of producing high quality films. Examples of maintenance and cleaning tasks include:

- draining and cleaning the tanks and rinsing thoroughly before adding the new chemicals
- cleaning the rollers with a soft sponge and mild detergent
- wiping down the rollers and feed tray daily

It is important to always turn on the processor prior to taking radiographs. This allows time for the processor to heat up to the correct temperature for the chemicals and for film drying. Most processors require the temperature to reach 95° F (35° C) and have built-in sensors that control the temperature. When the optimal temperature is reached, a light will come on, indicating that the processor is ready to accept films.

USE of an Automatic Processor

TECHNICAL ACTION	RATIONALE/AMPLIFICATION
1. Turn on processor and allow it to reach operating temperature.	
2. Turn on safelight and outer "in use" warning light.	
3. Turn off standard lights in the darkroom and allow your eyes to adjust to the safelight illumination.	
4. Open cassette and imprint film.	4a. Turn the cassette over and unlatch from the back.
	4b. Handle films only by the corners, and carefully insert the unexposed corner into the film imprinter to record patient information prior to developing.

(Continues)

TECHNICAL ACTION	RATIONALE/AMPLIFICATION
5. Align the edge of the film with the feeder tray and advance until the roller bar takes it.	5a. Do not advance the film by hand; always allow the roller bar to accept it and pull it forward into the processor.
	5b. As there are usually at least two views to be developed, wait until the processor signals that the first film has advanced far enough to insert another. Depending on the model, this may be indicated by a light or a short beep.
6. The developed and dried film will roll out from the opposite end of the feeder tray.	6a. Do not pull the films from the last roller, but allow the film to clear prior to touching it.
7. Reload cassettes with new film prior to opening the darkroom door or turning on any other light.	7a. Cassettes should always be ready and never be left without film
8. Once all films have been accepted into the processor and the used cassettes have been reloaded with film, it is safe to open the darkroom door.	

FAST FACTS

Dogs

Terminology

- ➤ **Male:** dog (entire, castrated, or neutered)
- ➤ **Female:** bitch (spayed)
- ➤ **Young:** pups or whelps
- ➤ **Adult weight:** breed dependent, 2 pounds to 110 pounds (variable)

Life Span

- ➤ breed dependent, 7–11 years (large dog); 15–18 years (small dog)

Reproduction

- ➤ dog sexually mature 8–10 months, bitch 5–11 months. Estrus cycle approximately every 6 months (average)

Gestation

- ➤ 63 days. Process of giving birth; whelping
- ➤ Litter size: small breeds average 2–4 pups, large breeds average 5–12 pups
- ➤ Weaning age: 6–8 weeks

Vaccinations

- ➤ Combination vaccine: Distemper, hepatitis (canine adenovirus) leptospirosis, parvovirus, parainfluenza
- ➤ Additional vaccines: (Bordatella bronchiseptica, corona virus, Lyme disease)

➤ Required: Rabies
➤ Heartworm Preventative

Vital Signs

➤ **Temperature:** 100–101° F. 37.5–39°C
➤ **Pulse (HR):** 70–160 beats/minute
➤ **Respiration:** 8–20 breaths/minute

Dental Formula

➤ Deciduous teeth: 2 × (3i/3i, 1c/1c,3p/2p) total 28
➤ Permanent teeth: 2 × (3I/3I,1C/1C. 4P/4P, 2M/3M) total 42
➤ **Zoonotic Potential** (Selected)
➤ **Bacterial**
  ➤ Leptospirosis
  ➤ Brucella (spp)
  ➤ Salmonella
➤ **Viral**
  ➤ Rabies
➤ **Fungal**
  ➤ Microsporum spp Ringworm
➤ **Protozoan**
  ➤ Giardia
➤ **Parasites**
  ➤ Toxocara spp: Roundworms, Anclyostoma; Hookworms (cutaneous larval migrans)
  ➤ Echinococcus spp: Tapeworm (Hydatid disease)
  ➤ Sarcoptes scabei; Scabies, "mange mite"

Cats

Terminology

➤ **Male:** tom (castrated or neutered)
➤ **Female:** queen (spayed)
➤ **Young:** kitten
➤ **Process of giving birth:** kindling
➤ **Life Span:** 14–20 years
➤ **Adult Weight:** breed variable, average 8–16 pounds

Reproduction

➤ **Tom:** sexually mature 7–9 months
➤ **Queen:** sexually mature 5–7 months, induced ovulator, seasonally polyestrus
➤ **Gestation:** 63 days
➤ **Litter size:** 3–6
➤ **Weaning age:** 6–8 weeks

Vaccinations

- Panleukopenia (feline distemper) Chlamydia pneumonitis, Rhinotracheitis, Calici virus, Feline infectious peritonitis (FIP) Feline immunodeficiency virus, Feline Leukemia
- Rabies (recommended but not required in many states)

Vital Signs

- **Temperature:** 101–102° F, 38–39° C
- **Pulse (HR):** 150–210 (beats per minute)
- **Respiration Rate:** 8–30 (breaths per minute)

Dental Formula

- **Deciduous teeth:** $2 \times (3i3i, 1c/1c/3p/2p) = 26$
- **Permanent Teeth:** $2 \times (3I/3I, 1C/1C, 3P/2P, 1M/1M) = 30$
- **Zoonotic Potential:** (selected)
- **Bacterial**
  - Bartonella henselae (Cat Scratch Fever)
  - Leptospirosis
  - Salmonella
  - Pasteurella
  - Yersinina pestis (Plague)
- **Viral**
  - Rabies
- **Fungal**
  - Microsporum spp (Ringworm)
- **Protozoan**
  - Toxoplasmosis
  - Giardia
- **Parasites**
  - toxocara, spp (Roundworms)
  - Anclostoma spp (Hookworms)
  - Dipylidium spp (tapeworm)
  - Echinoccus (tapeworm)

Review Questions

1. What are four methods of restraint that can be used with a cat?

2. Name three factors that determine which type of restraint should be used with a dog.

3. What is the major difference between a cat muzzle and a dog muzzle?

4. List the veins most frequently used for blood collection.

5. What is the difference between a purple top tube and a red top tube?

6. Why should pregnant women avoid cleaning a cat litter box?

7. A disease that is transmitted from animals to humans is called a _____

8. Calculate the kilogram weight of a dog that weighs 38 pounds.

9. How is the correct milligram dose of a medication calculated?

10. What is the difference between IV fluids described as hypertonic or hypotonic?

11. List three observable signs of dehydration.

12. If a patient is to receive 1,500 ml of fluids over a 24-hour period, calculate the hourly drip rate.

13. What is the purpose of the pop-off valve on an anesthesia machine?

14. Describe the importance of PPV (positive pressure ventilation).

15. List three signs in a patient that occur during the recovery stage of anesthesia.

16. At what stage during recovery should the patient be extubated?

17. What is the purpose of a dosimetry badge?

18. In radiology, describe how the Technique Chart is used.

19. List the personal safety equipment used in radiology.

20. What is the difference between radiolucent and radiopaque?

References

Blood, D.C., & Studdert, V.P., *Saunders Comprehensive Veterinary Dictionary*, 2nd ed. 1999, WB Saunders.

Crow, Steven E., & Walshaw, Sally O., *Manual of Clinical Procedures in the Dog, Cat & Rabbit*, 2nd ed. 1997, Lippincott Williams.

Hendrix, Charles M., & Sirous, M., *Laboratory Procedures for Veterinary Technicians*, 5th ed. 2007, Mosby.

McCurrin, Dennis M., & Bassert, Joanna M., *Clinical Textbook for Veterinary Technicians*, 5th ed. 2002, WB Saunders.

Rockett, Jody, & Bassert, Susanna, *Veterinary Clinical Procedures in Large Animal Practice*, 2007, Cengage Learning.

Sirois, Margi (Consultant), *Mosby's Veterinary PDQ*, 2009, Mosby Elsevier.

Sirois, Margi, *Principles and Practice of Veterinary Technology*, 2nd ed., 2004, Mosby

Warren, Dean M., *Small Animal Care & Management*, 3rd ed., 2010, Delmar Cengage Learning

http://www.cvmbs.colostate.edu (accessed July 18, 2012)

http://www.dvm360.com (accessed July 23, 2012)

http://www.anesthesia2000.com (accessed March 3, 2013)

http://www.catvets.com (accessed February 16, 2013)

http://www.arst.org (accessed February 8, 2013)

http://www.rpop.idea.org (accessed February 2, 2013)

http://www.loudoun.nvcc.edu (accessed February 2, 2013)

http://www.vet.purdue.edu (accessed February 6, 2013)

http://www.asevet.com (accessed February 6, 2013)

http://www. isfm.net (accessed February 6, 2013)

http://www.diagnostic imaging.com (accessed February 8, 2013)

For additional reading:

http://www.mcalc.com (accessed July 16, 2012) (Nursing math)

http://www.dummies.com (accessed July 16, 2012) "Medical Calculations for Dummies Cheat Sheet"

http://www.vetmed.wsu.edu (accessed August 30, 2012)

CHAPTER 2

RABBITS

He didn't mind how he looked to other people, because the nursery magic had made him Real, and when you are Real shabbiness doesn't matter.

— Margery Williams, *The Velveteen Rabbit*

UNIT ONE

Overview of Species

OBJECTIVES

Upon completion of this unit, the reader should be able to:

❱ Describe basic rabbit behaviors

❱ Determine the sex of a rabbit

KEY TERMS

altrical

buck

cecotrophs

coprophagic

crepuscular

dewlap

doe

inguinal

kindle

kits

submandibular

RABBITS

Rabbits have become popular household pets partly because of the changing lifestyle of pet owners (**Figure 2-1**). For example, people may not find it practical to have a dog or a cat if they work all day or live in an apartment. Rabbits are **crepuscular**, which means that they are more active at dawn and dusk and spend much of the day taking short naps. They become active during the early evenings when the owners are home to enjoy their company and care for them. Clients often say that they are more comfortable leaving their pet rabbits during the day and do not feel as guilty as they would about leaving a dog or a cat.

▲ **FIGURE 2-1** Rabbits have become a popular household animal companion.

Other reasons for the rising popularity of "house rabbits" are that they are highly intelligent and will engage in play with their owners by readily chasing balls and objects tossed into the air. Rabbits normally urinate and defecate in one area and will use a litter box. They are easy to feed, drink water from a sipper bottle, and require simple housing.

BEHAVIOR

In general, rabbits are quiet animals and vocalize only when provoked, when in pain, or when extremely frightened. Growling is unmistakable, aggressive, and not just an empty threat. Rabbits will charge, growling and clawing with their front feet. The scream of a rabbit in pain is loud and piercing, a sound that is unforgettable to those who have heard it.

Although they can be affectionate and playful when socialized at an early age, rabbits can be very territorial. They will not hesitate to confront an intruder, whether it is a child, a visitor, or another household pet. Territorial aggression is common in rabbits of both sexes, but especially those that have not been spayed or castrated. Typically, an unspayed female, a **doe**, is more aggressive than an unneutered male, a **buck**.

Rabbits use scent glands to mark territory. They have **submandibular** or *chin glands*, and both does and bucks have **inguinal** glands located within the folds of skin on either sides of the genitals. Inguinal glands normally produce an unpleasant smelling, thick, and somewhat waxy substance.

Rabbits produce two types of fecal pellets. Hard pellets are passed during the day and consist of undigested waste. Night feces, **cecotrophs**, are soft and encased in a mucous membrane. Cecotrophs are consumed by the rabbit directly from the anus. Consumption of cecotrophs is essential to the rabbit's health, as they contain quantities of vitamins B and K, beneficial bacteria and protein. **Coprophagic** is a term that describes a species that normally consumes its feces. It is not only normal in rabbits but also in many rodent species.

BREEDING

Sexing rabbits can be a little more difficult, especially with young **kits**. The male's testicles descend around 12 weeks of age (**Figure 2-2**). This is complicated further because the inguinal canal remains open, allowing the testicles to retract into the abdominal cavity. Bucks have a rounded urethral opening

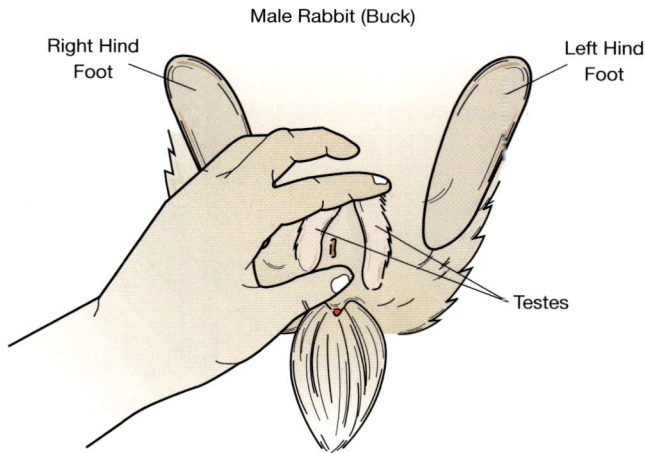

▲ **FIGURE 2-2** External genitalia of a buck.

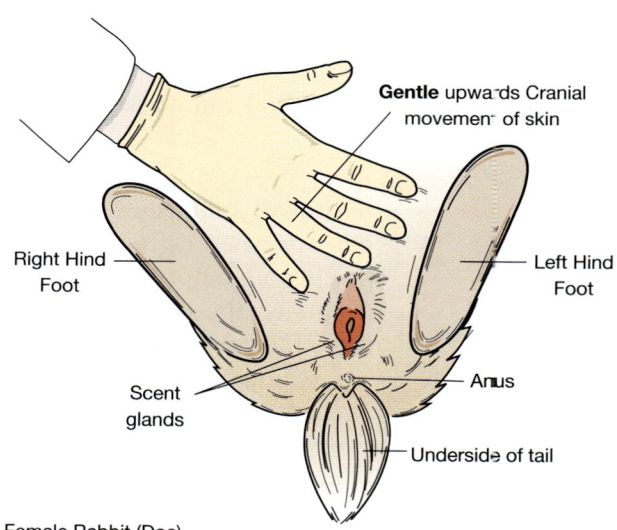

▲ **FIGURE 2-3** External genitalia of a doe.

▲ **FIGURE 2-4** The mature doe in many breeds has a pronounced dewlap.

and a sheath that covers a relatively small penis. Does have an elongated vulva with a vertical opening (**Figure 2-3**). With maturity, the does in many breeds may have a pronounced **dewlap** (**Figure 2-4**). The dewlap is rarely seen in bucks.

Does become sexually mature around 4–8 months—approximately 1–2 months earlier than bucks. Sexual maturity is influenced by breed and, as a general rule, smaller breeds mature earlier than larger breeds.

Does are induced ovulators, capable of breeding any time after they reach sexual maturity. Ovulation occurs within 10 hours after breeding. Because of the territorial nature of the female, the doe should be taken to the buck for breeding and returned to her own cage after mating. The presence of a buck may be stressful to the doe and may be one cause of reabsorption of the embryos.

Bucks have been known to kill new litters, and does sometimes cannibalize the young. Once a doe starts to cannibalize her young, she is likely to continue doing this with each litter. Does also consume dead or deformed kits.

Gestation is 29–35 days. Most does give birth, or **kindle**, in the early morning. The young are born **altrical**, meaning that they are born hairless with their eyes closed and ears sealed. Many rabbit owners become concerned when they do not see the doe nursing her kits or in the nest with them. Owners should be advised not to remove any kits from the nest in an attempt to bottle-feed. Does normally nurse their young only 3–5 minutes per day. The kits nurse on their backs and consume approximately 20% of their body weight per feeding. Rabbit milk is high in fat and difficult to replicate. Attempting to feed kits from a bottle, even on their backs, often results in aspiration. Orphaned kits are difficult to rear to weaning age.

UNIT TWO

Restraint Techniques

OBJECTIVES

Upon completion of this unit, the reader should be able to:

▶ Correctly restrain a rabbit for various procedures

KEY TERMS

dyspnea

GUIDELINES FOR RESTRAINT OF THE RABBIT

Rabbits, while popular as pets, are still considered exotic because they are different in anatomy, physiology, and behavior from more traditional companion animals. The veterinary technician must become familiar with these differences, and it is essential to be able to restrain the rabbit correctly for various clinical procedures.

The most important thing to remember when handling or restraining a rabbit is how easily the patient can be injured. The skeleton of a rabbit is comparatively fragile, and the well-muscled hind limbs have a powerful kick. This, coupled with the long, but relatively inflexible spinal column and the force of a rabbit's kick can be enough to fracture its back. The hind limbs should always be securely restrained and the body supported in the normal curved posture. A spinal fracture caused by kicking out in an attempt to escape from restraint usually occurs between lumbar vertebrae L6/L7. The result is paralysis and consequent euthanasia. *A rabbit should never be picked up by just a scruff or by the ears.*

PURPOSE

- To restrain and control the rabbit for an examination
- To administer medication and injections
- To perform nail and teeth trims
- To induce anesthesia

COMPLICATIONS

- Scratches to personnel
- Fracture of patient's lumbar vertebrae
- Hyperthermia
- Stress and difficulty breathing, termed **dyspnea**

EQUIPMENT

- Heavy towels

PROCEDURE for Removing a Rabbit from a Cage and Returning a Rabbit to a Cage

TECHNICAL ACTION	RATIONALE/AMPLIFICATION
1. Stand quietly in front of the cage and observe the rabbit's attitude.	1a. If the patient is already stressed or has respiratory compromise, restraint could increase dypsnea. Observe the rate and depth of respirations.
2. Talk to the rabbit and quietly open the cage door with your body blocking the doorway.	2a. 3a. 4a. Rabbits are a prey species. The close approach of a strange person in a strange environment may be perceived by the rabbit as a predator and one that has it trapped! If a rabbit is alarmed, it will thump loudly with a hind foot. The "cornered" rabbit will either attempt to elude capture by hopping away or will become aggressive and attack. The eyesight of a rabbit is poor, the eyes are placed laterally, and it has very little depth perception.
3. Attempt to stroke the rabbit from the back of the head.	
4. If the patient demonstrates alarm or aggression, toss one towel into the cage and another quickly over the top of the rabbit.	

(Continues)

TECHNICAL ACTION	RATIONALE/AMPLIFICATION
5. Scruff the rabbit with one hand, and use the other hand to support the hind quarters with a firm grasp on both hind legs. 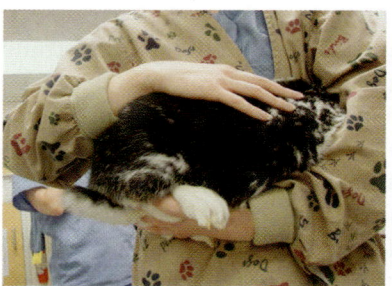 ▲ **FIGURE 2-5** Secure the rabbit's hind legs, and hold the rabbit close to your body so that its spine is in a natural curved position.	5a. Docile rabbits may be picked up without the use of a towel. Remember: "SSF" (Scruff, Seat, Feet). When holding the rabbit, keep its back curved to prevent injury to the spinal column (**Figure 2-5**).
6. Remove from cage and tuck the rabbit's front feet and head under your arm.	6a. The rabbit should be held close to the restrainer's body. When the head is tucked under the arm, the rabbit feels more secure.
7. When returning the rabbit to its cage, always place it in backwards, hind end in first.	7a. Placing the rabbit in the cage hind end first reduces the chances of its kicking-out during the release from restraint and, from the rabbit's point of view, "escape." To go anywhere, it must turn around first. More often than not, the released rabbit will "freeze," then quietly turn and hop to the rear of the cage. However, rabbits can be unpredictable, so always be prepared and do not assume that a behavior is going to be "textbook."

PROCEDURE for Towel Restraint

TECHNICAL ACTION	RATIONALE/AMPLIFICATION
1. Maintain the scruff and place the rabbit in the center of a large towel. 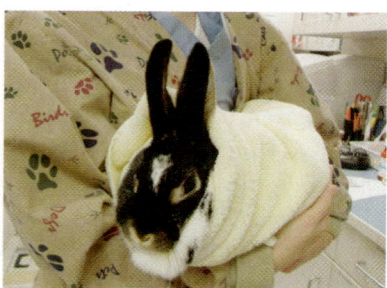 ▲ **FIGURE 2-6** The rabbit may also be wrapped in a towel, creating a "bunny burrito."	1a. The towel wrap, also called the "bunny burrito," is useful for a variety of procedures. Even though the rabbit is wrapped securely in the towel, a restrainer always should have his or her hands on the rabbit in a towel, facing away from the restrainer's body. The restrainer should place his or her body directly behind and close to the rear end of the "burrito," with the hands placed around the thorax (**Figure 2-6**).

(Continues)

TECHNICAL ACTION	RATIONALE/AMPLIFICATION
2. Wrap one side of the towel securely around the rabbit's neck and shoulders.	**2a.** The towel should enclose at least one front foot. Rabbits use both front feet together when they hop.
3. Release the scruff and wrap the other side of the towel around the rabbit's body.	
4. The head, including the ears, should be uncovered.	**4a.** Rabbits do not have sweat glands, and their ears function in thermoregulation, so the ears should not be folded back and wrapped in the towel. Observe the patient carefully for signs of stress and hyperthermia by watching the nares for increased breathing and feeling an increase in respiratory effort from the thorax. These are both signs that the patient is stressed and becoming overheated. If this occurs, release the towel wrap immediately and restrain the rabbit by the scruff and hind quarters. It may be necessary to release the rabbit onto the floor and, as always, facing the handler and hind end first.

RABBIT HYPNOSIS

A rabbit can be put into a trance-like state by covering its eyes and slowly turning it onto its back (**Figure 2-7**). With the rabbit in dorsal recumbency, stroke the full length of the abdomen, moving the hand caudally in the direction of fur growth. Small gentle, circular movements of the hand also work well. The rabbit will relax completely, making an examination of the ventral area easy. Do not use this technique for any procedure that is likely to cause pain. The rabbit is easily roused by returning it back to sternal recumbency. Do so with a scruff and restraint of the hind legs.

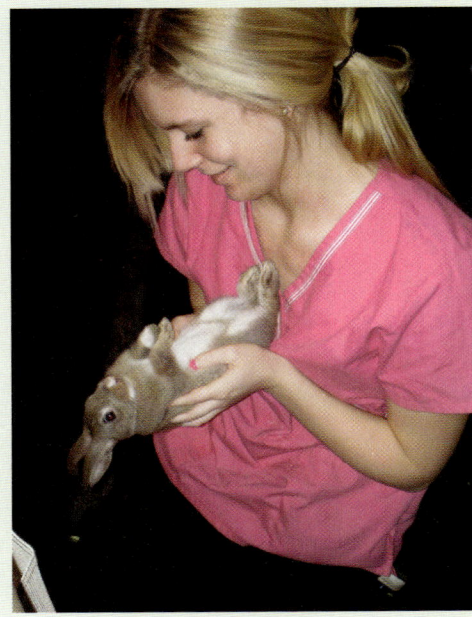

◄ **FIGURE 2-7** Quietly placing the rabbit in dorsal recumbency and rubbing the abdomen will induce a trance-like state.

Grooming

OBJECTIVES

Upon completion of this unit, the reader should be able to:

▶ Perform nail and teeth trims and basic grooming

KEY TERMS

malocclusion

peg teeth

© djem/Shutterstock.com

GUIDELINES FOR GROOMING THE RABBIT

There are many varieties of rabbits, from the long-haired Angora to the velvety-coated short-haired Rex. Although rabbits groom themselves similar to that of a cat, they all benefit from regular grooming with a soft brush. This helps remove loose hair and promotes healthy skin. Rabbits ingest hair and often suffer from hairball impaction.

Routine grooming procedures for the veterinary technician include trimming the nails, clipping or reducing the length of overgrown incisor teeth, and perhaps giving a bath to a cage-soiled rabbit. To perform any of these tasks, veterinary technicians must become competent in rabbit restraint and work as a team.

GUIDELINES FOR TRIMMING THE NAILS OF A RABBIT

Nail trims for the rabbit are no different from the procedure performed in other small animals except that the need for careful approach and restraint is greater because of the potential of injury to the patient.

PURPOSE

- To trim nails to the correct length, making the rabbit more comfortable and easier for the owner to handle
- To prevent such severe overgrowth of a rabbit's nails that normal locomotion is inhibited
- To prevent the nails from becoming entangled in carpets and clothing

COMPLICATIONS

- Stress to the patient during restraint
- Hyperthermia

EQUIPMENT

- Any type of small-animal nail trimmers
- Assistant for restraint
- Towel for restraint if rabbit is resistant or aggressive
- Styptic powder

PROCEDURE for a Rabbit Nail Trim

TECHNICAL ACTION	RATIONALE/AMPLIFICATION
1. Have the restrainer place the rabbit on a solid surface.	1a. Depending on the temperament of the rabbit, towel restraint may not be necessary.
	1b. The restrainer should tuck the rear end of the rabbit close to his or her body to prevent the rabbit from kicking out. The head is facing the person performing the trims (**Figure 2-8**).

(Continues)

TECHNICAL ACTION	RATIONALE/AMPLIFICATION

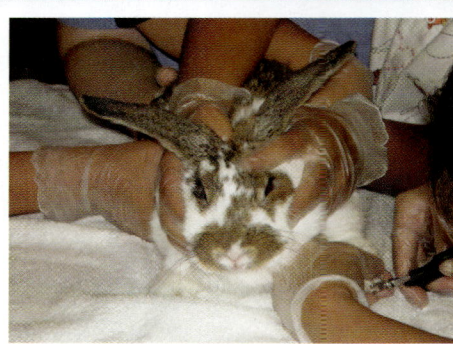

▲ **FIGURE 2-8** To trim the rabbit's nails, hold its rear end close to your body with its head facing the technician doing the trimming.

2. Gently pull one foot forward and separate the fur away from each nail.

2a. It is easier and quicker to trim front and back nails on one side, then move across to the other side.

2b. Rabbits have five toes on each foot. On the forefeet, one claw is a declaw, located higher on the medial/lateral (inside) of the leg.

2c. Rabbits do not have footpads but a dense covering of fur where the footpads in other species are located. The fur is especially thick on the hind feet.

3. Use nail clipper to cut back each nail to the appropriate length, being careful not to cut into the quick.

4. If cut too short, apply styptic powder to stop the bleeding.

5. Continue the procedure by trimming front and rear nails on one side, then move to the other side.

GUIDELINES FOR CLIPPING THE INCISOR TEETH

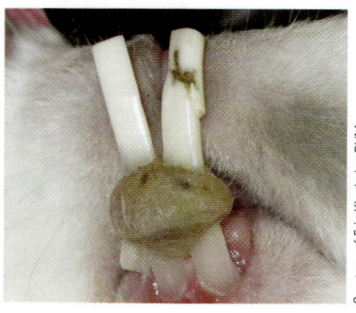

Courtesy of Eric Klaphake, DVM

▲ **FIGURE 2-9** Extremely overgrown lower incisor teeth is a result of malocclusion; here, the left incisor is fractured and a large hairball has formed around both upper incisors.

The most common problem seen in rabbits is overgrowth of the incisors. Rabbit teeth are open-rooted and grow continuously. **Malocclusion**, poor alignment of the top and bottom incisors, prevents the teeth from being worn down normally. When this condition is present, the lower incisors tend to grow up and out and the upper incisors tend to curl back into the oral cavity (**Figure 2-9**).

Overgrown teeth may be trimmed to the correct length with small wire nippers or a hand-held rotary tool with a cutting wheel. Rabbits have two sets of top incisors, one behind the other. The second pair of incisor teeth, called **peg teeth**, usually do not have to be clipped back and often are found to be somewhat loose (**Figure 2-10**).

PURPOSE

- To allow the rabbit to eat normally
- To prevent the top incisors from curling into the oral cavity

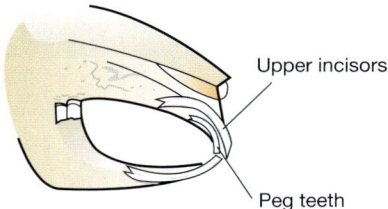

▲ **FIGURE 2-10** Rabbits have two sets of upper incisors; the peg teeth are located directly behind the primary incisors.

- To prevent overgrown incisors from fracturing into the tooth socket
- To prevent oral ulcers

COMPLICATIONS

- Dull or chipped cutting surfaces on the wire cutters may cause damage to the teeth
- Cutting lips or tongue with rotary tool blade
- Embedded upper incisors
- Existing fractures to the incisors
- Existing sores and ulcers
- Splintering of tooth when clipped
- Small oral cavity of the patient
- Stress and hyperthermia

EQUIPMENT

- Safety eyeglasses
- Pair of small wire cutters
- Hand-held rotary tool with cutting wheel
- Small pair of blunt forceps
- Dental mirror
- Assistant for restraint
- Towel for restraint

PROCEDURE for Clipping Incisor Teeth

TECHNICAL ACTION	RATIONALE/AMPLIFICATION
1. Put on safety eye glasses.	1a. The teeth are fairly brittle and often become projectiles when clipped. Protect your eyesight!
2. The rabbit should be on a solid table. (For restraint method, refer to Figure 2-8 and 1a.1b., Nail Trim restraint.)	
3. Put one hand over the top of the rabbit's head, avoiding pressure on the eyes.	
4. Hold the wire cutters or rotary tool in the other hand, so that when cutting, the approach to the tooth is horizontal to the incisor.	4a. Great care must be taken if using a hand-held rotary tool and cutting blade. The action is quick and if control is lost, it may cause severe cutting injuries to the mouth and tongue.
5. Raise the upper lips with the thumb and forefingers of the hand holding the head.	
6. Inspect incisors and determine where to cut.	
7. Make a decisive snip through one of the lower incisors; repeat for the other lower incisor.	
8. Trim both incisors evenly.	

(Continues)

TECHNICAL ACTION	RATIONALE/AMPLIFICATION
9. Repeat steps 5–8 for the upper incisors.	
10. Remove any sharp points with a hand-held dental file or with the use of a rotary tool.	
11. Re-inspect teeth for fractures, and check for any oral bleeding.	
12. Examine the roof of the mouth with the use of a dental mirror.	12a. The oral cavity of a rabbit is very small, so the mouth cannot be opened wide enough to inspect the roof of the mouth. Use the dental mirror to look for imbedded pieces of the upper incisors and oral lesions.

Motorized dental burs and small cutting disks are also used for teeth trimming. When using a motorized device, injury including severe cuts to the lips and tongue can occur quickly if the rabbit is not completely immobilized. Sedation or anesthesia may be required when using this method. In addition, a motorized dremel-type tool produces dental dust, which could be inhaled by the patient or lodge in the eyes. These tools also produce heat, so thermal burns to the mouth and tongue can occur. This problem must be carefully guarded against.

Rabbit teeth grow continually, and trimming may be needed every 6–8 weeks. The additional cost of anesthesia or sedation for each visit is prohibitive to many clients, who prefer the standard "clipping" of incisor teeth, which takes only a few moments and there is no recovery time.

GUIDELINES FOR GIVING A RABBIT A BATH

As strange as it may seem, many rabbits enjoy water. The toes of the hind feet are webbed, and they are capable swimmers. Some rabbits willingly hop into a tub of shallow water, apparently to cool off. It is not recommended to provide a rabbit with a pool, but there are countless reports of rabbits intentionally entering water, even joining the family in their outdoor pool. Giving a rabbit a bath is a lot less harrowing than giving a bath to a cat!

The water should be cool but not cold. Any non-medicated kitten shampoo can be used. Because of the density of rabbit fur, great care should be taken to remove all soap residue. Do not hold the rabbit directly under running water but, rather, use a basin to pour the water over the rabbit. If using a spray nozzle, the stream should be gentle, not forceful, and it can be held directly onto the fur, moving it around similar to a massage. Avoid getting any soap or water into the rabbit's eyes or inside of the ears.

If the rabbit is presented with signs of mild heat stroke, it can be rapidly cooled down by placing it in a tub of cool water. If this is the case, do not use any shampoo, but continually support the patient while pouring the cool water over the patient.

PURPOSE

- To bathe a cage-soiled rabbit
- To cool down an over-heated rabbit

COMPLICATIONS

- None

EQUIPMENT

- Use the standard stainless steel treatment room wet table tub
- Several large towels
- Rubber mat
- Kitten shampoo or hypoallergenic shampoo
- Basin

PROCEDURE for Giving a Rabbit a Bath

TECHNICAL ACTION	RATIONALE/AMPLIFICATION
1. Remove rack from the wet tub in the treatment area.	1a. Do not take the rabbit into the grooming room used for dogs. The scent of dog hair (predator) will cause a great deal of stress for the rabbit.
	1b. Wet table tubs are deep, straight-sided, and made of stainless steel, which makes it easier to remove the scent of other animals. The rabbit is less likely to attempt to jump out or scramble up the sides of the wet table as it may do if using a normal (porcelain) bath tub.
2. Rinse tub completely, including all sides, and clear the drain.	2a. This is to remove any residual hair, debris, or cleaning agents from the tub.
3. Place the rubber mat on the bottom of the tub.	3a. The rubber mat prevents the rabbit from slipping and provides some traction.
4. Put in the drain plug.	
5. Run cool water to a depth of about 4–6 inches, depending on the size of the rabbit.	
6. Collect the patient and restrain in a "bunny burrito."	
7. Gently lower the rabbit into the tub of water and unwrap the towel, keeping one hand on the lower back until you see how the rabbit is going to react.	7a. This is to keep the rabbit safely restrained instead of just lowering it straight into the water. Once the towel is removed, the rabbit will likely sit still in the water.
8. Pour water over the rabbit with the basin until the fur is thoroughly wet or, if using the spray nozzle, turn on the water slowly with the head of the sprayer under the water.	8a. The sudden sound of the water from a spray nozzle is likely to startle the rabbit; if it is turned on under the water it is less likely to startle the rabbit.
9. Add shampoo, and proceed as you would with a small dog.	
10. Prior to rinsing, pull the tub drain and ensure that part of the rubber mat partially covers the drain.	10a. Allow the water to drain from the tub so that adding more water for the rinsing doesn't make it too deep. The rubber mat prevents the rabbit's nails from getting caught in the drain.

(Continues)

TECHNICAL ACTION	RATIONALE/AMPLIFICATION
11. When all the soap has been removed, gently "squeegee" the fur, especially the dense fur on the hind feet, to remove as much water as possible.	
12. Begin to towel dry with rabbit still in the tub.	
13. When most of the water has been removed by the towels, place the rabbit in a dry towel and wrap it in another "burrito" for removal from the tub.	
14. Place the rabbit on the floor or have an assistant receive it before releasing the rabbit from the towel wrap or returning it to its cage.	14a. Towel drying can continue on the floor. The person should sit with the rabbit nestled between the legs, head facing away. If the rabbit is going to be returned to the cage, the cage floor should be covered in towels, not newspaper, because newsprint will come off onto the clean, damp fur, making it black and towels are more absorbent than newspaper. The rabbit will begin to groom itself, further drying and fluffing the fur. No matter how closely supervised, rabbits should never be put in a dryer cage because of the danger of hyperthermia and the frightening sound of forced air.

Patient History

© djem/Shutterstock.com

OBJECTIVES

Upon completion of this unit, the reader should be able to:

- Take a complete patient history
- Discuss the importance of diet and digestion in the rabbit

KEY TERMS

cecum	hutch
gut-motility	urolith

GUIDELINES FOR TAKING THE PATIENT HISTORY

The patient history is similar to that of any other species, and the standard small-animal form is used. Begin with the chief complaint, complete the signalment, and continue gathering information with regard to housing, diet, and any previous medical problems. During this time, the rabbit should be left in its carrier.

PURPOSE

- To obtain as much information as possible in order to complete the physical exam and prepare for the veterinarian's consultation.

COMPLICATIONS

- None

EQUIPMENT

- Patient History form in computer file
- Or hard copy of Patient History form
- Pen

PROCEDURE for Taking a Patient History for the Rabbit

TECHNICAL ACTION	RATIONALE/AMPLIFICATION
1. Begin with the chief complaint, the reason for the visit.	
2. Complete the signalment.	2a. Many times clients do not know the sex of the rabbit, especially if it is a young kit. This can be determined during the physical examination and recorded then.
3. Methodically proceed, following the questions on the form. (Vaccination status is not asked for rabbits.)	3a. There are no recommended vaccinations for rabbits. Because no area of the exam form should be left blank, enter "n/a" (not applicable) regarding vaccination status.
4. Note how the rabbit was acquired and if there are any other household pets.	4a. Rabbits are a prey species, so the presence of a ferret or a dog in the home could be a major source of stress to the rabbit.
5. Record as much information on the rabbit's diet as possible, including supplements, treats, and items provided for chewing (natural non-toxic wood, flavored chew blocks, alfalfa blocks, etc.).	5a. Feeding the correct diet is critical to a rabbit's health. It should consist of formulated pellets and grass hay supplemented with fresh leafy greens.
	5b. Formulated rabbit pellets contain analyzed amounts of crude protein and fiber. Pellets that are high in fiber help reduce hairball formation and stimulate **gut-motility**, the normal way food is moved through the digestive tract. High-protein diets are not recommended for extended periods of time because a high-protein/low-fiber diet increases the potential for diarrhea and **urolith** (bladder stone) formation (**Figure 2-11**).

(Continues)

TECHNICAL ACTION	RATIONALE/AMPLIFICATION

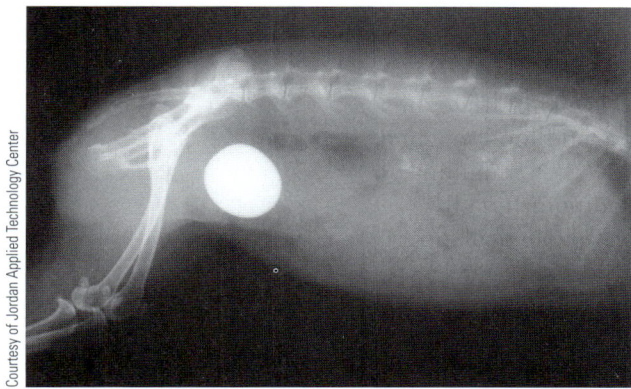

▲ **FIGURE 2-11** This large urolith (bladder stone) was discovered during a routine wellness exam.

5c. Rabbits are monogastric, hind-gut fermenters. What this means is that they are single-stomached, with microbes in the **cecum** that ferment (convert) organic material into usable carbohydrates. The cecum is part of the large intestine and one of the largest organs in the rabbit's body. In this way, a rabbit is more similar to a horse than to a dog or a cat (**Figure 2-12**).

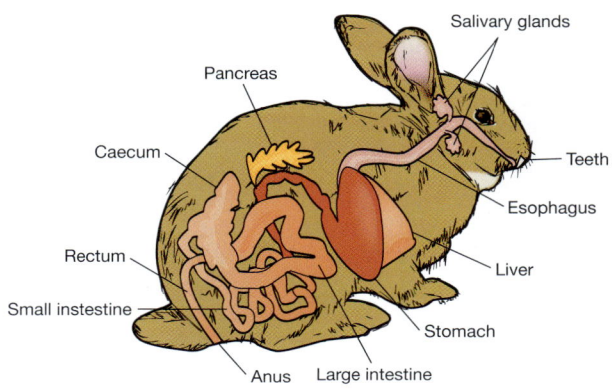

▲ **FIGURE 2-12** Gastrointestinal system of a rabbit.

6. Ask the client to describe the habitat in detail.

6a. Rabbits can be kept outside in a **hutch** (a cage specifically built for rabbits), indoors, or a combination of both. All habitats should have an area of solid floor to get off the wire, and a hide-box.

6b. Rabbits kept outdoors need a more thorough examination of their fur. (Refer to Procedure for a Physical Examination of the Rabbit.)

7. Record any details of previous medical problems.

Physical Examination

OBJECTIVES

Upon completion of this unit, the reader should be able to:

▶ Identify the steps in the physical examination of the rabbit

▶ Assist in the examination as required

KEY TERMS

hutch burn	ringworm
papillomas	Snuffles
Pasteurella	spirochete
pododermatitis	torticollis
Rhinitis	

GUIDELINES FOR THE PHYSICAL EXAMINATION OF THE RABBIT

Performing a physical examination of the rabbit can be a little more challenging than for that of a small dog or a cat. Great care must be taken in handling the patient, and it is often a good idea to have an assistant for restraint because of the potential of injury to a rabbit's back if it becomes frightened and attempts to escape. In rabbits, a disease may affect multiple areas of the body, so it is important to recognize various signs discovered during the physical examination that are linked and may lead to a direct diagnosis by the veterinarian.

PURPOSE

- To record objective information
- To assess subjective information
- To assist the veterinarian in his or her own evaluation of the patient

COMPLICATIONS

- Restraint considerations
- Temperament of the rabbit

EQUIPMENT

- Small mammal examination form, computer or hard copy
- If using hard copy, pen and clipboard
- Stethoscope
- Thermometer
- Gram scale
- Towel
- Assistant for restraint
- Disposable gloves

PROCEDURE for the Physical Examination of the Rabbit

TECHNICAL ACTION	RATIONALE/AMPLIFICATION
1. Place a large towel on the examination table.	
2. Wear disposable gloves.	
3. Remove patient from carrier and place on a towel or directly on the gram scale to obtain weight.	
4. An assistant should control the rabbit's movements and be prepared for immediate restraint if necessary.	
5. Observe quietly for respiration rate, and record.	5a. Average respiration rate for a rabbit is 35–60 breaths/minute.
6. Auscultate for the heart rate, and record.	6a. The heart rate for a rabbit can vary considerably, from 135–325 beats/minute, and still be within normal limits. The immediate environment should be taken into consideration, and it is best to obtain a heart rate prior to proceeding with the rest of the exam.

(Continues)

TECHNICAL ACTION	RATIONALE/AMPLIFICATION

7. Auscultate the lungs.

8. Examine all areas of the head—eyes, nostrils, dewlap, teeth, and ears—and for the presence of a head tilt.

8a. The eyes should bright and be free of discharge. **Pasteurella** (spp.) is a common bacterial infection of rabbits that often affects the conjunctiva, and the resulting pus can completely occlude the lacrimal duct (**Figure 2-13**).

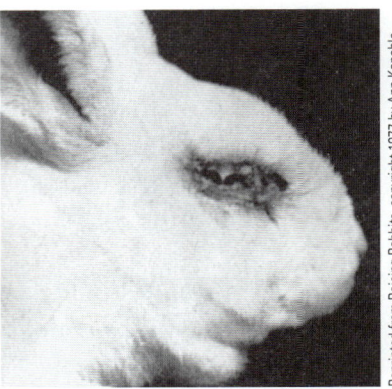

Reprinted from Raising Rabbits, copyright 1977 by Ann Kanable. Permission granted by Rodale Press, Inc., Emmaus, PA 18098

▲ **FIGURE 2-13** This rabbit is suffering from an inflammation of the conjunctiva of the eye ("weepy eye"); note the discharge from the eye and matting of the fur around the eye.

PASTEURELLOSIS

Pasteurellosis is one of the most common bacterial infections of rabbits. It affects not only the conjunctiva but also the entire respiratory system or any organ of the body. Signs of pasteurellosis are variable and can include nasal and ocular discharge, genital swelling, dermal ulcerations, weight loss, and sudden death. Asymptomatic carriers can harbor the bacteria in their reproductive organs, ears, nares, lungs, and conjunctiva.

TECHNICAL ACTION	RATIONALE/AMPLIFICATION

8b. Rabbits cannot breathe through the mouth. Observation of the rate of movement of the nares can help determine difficulty breathing or an abnormally high respiration rate. There should be no nasal discharge. **Rhinitis**, or "**Snuffles**," is caused by a pasturella infection. It is characterized by a thick discharge that is yellow or white. It adheres to the nares, and as the rabbit attempts to groom itself, it will stick to the forepaws and may spread to other parts of the body.

(Continues)

TECHNICAL ACTION	RATIONALE/AMPLIFICATION

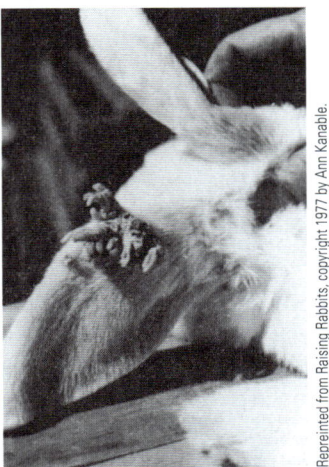

▲ **FIGURE 2-14** Papillomas on the ear of a rabbit. Papillomas are usually benign warts.

Reprinted from Raising Rabbits, copyright 1977 by Ann Kanable. Permission granted by Rodale Press, Inc., Emmaus, PA 18098

8c. The dewlap in does can become wet and slimy, either from dragging it through a water bowl (a change to a water sipper bottle is recommended) or from a constant dribble of saliva. If this is the case, it usually is the result of malocclusion and overgrowth of the incisor teeth.

8d. Examine the front incisors for malocclusion and overgrowth.

8e. Examine the ears gently, as they are extremely sensitive. Rabbits do not have sweat glands, and the ears assist in thermoregulation. Examine the inner ears for the presence of mites. Rabbits sometimes present with small growths on the outer ears that may spread to the cranial aspect of the neck. These **papillomas** usually are benign warts (**Figure 2-14**).

8f. A head tilt in the rabbit may be a serious condition called **torticollis**, caused by a bacterial infection that affects the nerves of the neck. The rabbit cannot straighten its head, and this prevents it from eating or drinking (**Figure 2-15**). Treatment is not always successful, and euthanasia is a common result of this condition.

Courtesy of Eric Klaphake, DVM

▲ **FIGURE 2-15** Torticollis, or wry neck affects the nerves of the neck, which prevents the rabbit from eating and drinking.

9. Look carefully at the fur and examine the skin underneath to see if there is any abnormality.

10. Gently palpate the entire body.

9a. 10a. Rabbits that are housed outdoors can be infested with maggots and/or cuterebra larvae. A rabbit that presents with a maggot infestation will have a distinctive foul odor, visible signs of maggot "holes" in the skin, and skin decomposition. Cuterebra are large flies, and the larvae burrow under the skin, creating small cyst-like cavities. There may be several maggot holes (with the larvae inside and visibly moving.) Although much larger in size, the cuterebra larva may be more difficult to visualize without carefully shaving the fur away from the suspect area to reveal a small breathing hole on the surface of the lump. Only one or several cuterebra larvae may be present, but each will form its own cyst-like cavity. Great care should be taken *not* to crush the living larvae of either

(Continues)

TECHNICAL ACTION	RATIONALE/AMPLIFICATION
	while still in the rabbit. *Do not squeeze any suspect bump/lump or abnormality.* Crushing the larvae *in situ* will cause a severe anaphylactic response in the rabbit and immediate medical, potentially life-saving, treatment will be necessary. Cuterebra larvae are frequently found on the underbelly. A maggot infestation is often near the vent or on the back, near the rump and tail, but all areas of the body should be examined.
	9b. 10b. Ringworm, a fungal infection, can affect rabbits that are kept outdoors. Look for round areas of alopecia or hair loss. The lesions are red and slightly raised. Ringworm is spread through direct contact. It is zoonotic (transferable to humans) and because of this potential, all personnel handling a rabbit should wear disposable gloves and wash their hands thoroughly while in the examination room. If ringworm is suspected, the lesion can be examined under a Wood's Lamp (black light) and it will fluoresce bright green. For a positive diagnosis, scrapings from the lesion and surrounding hair should be cultured.
11. Using the "trance technique," place the rabbit on its back to examine the vent area and confirm its sex.	**11a.** Rabbit syphilis, or "vent disease," is caused by a **spirochete**, a type of bacterium, and is transmitted from rabbit to rabbit via direct genital contact in infected breeding colonies. Rabbits with this disease can be asymptomatic until they are stressed. Lesions appear on the genitalia, perineum, eyelids, and mouth. Rabbit syphilis is *not* zoonotic.
	11b. Hutch burn, or urine scald, affects the external genitalia. It occurs when rabbits sit in soiled cages. The genitalia appear red and swollen, similar to a rash.
	11c. Rabbit urine is normally thick and creamy, but diet affects the consistency and color. Orange- or rust-colored urine alarms many owners, creating concern that there is blood in the urine. In most instances, the color indicates a diet high in protein. Clear urine may indicate that the patient has a higher calcium requirement.
12. Examine the hocks and bottoms of all four feet.	**12a.** Rabbits do not have footpads but, instead, a thick covering of fur on the bottom of their feet. When rabbits are kept in a hutch with only a wire floor, the hair can become matted, soiled, and rubbed away. The result is **pododermatitis**, "sore hocks" on the plantar surface of the hind feet.
13. Obtain a rectal temperature.	**13a.** The rabbit should be restrained on the table and the head tucked under the elbow like a football. Lubricate the thermometer tip and insert gently into the rectum. The temperature of a rabbit is between 38.3°–40° C (101°–104° F).

(Continues)

TECHNICAL ACTION	RATIONALE/AMPLIFICATION
14. Return the rabbit to sternal recumbency and allow it to settle for a few moments.	
15. Using the towel on the table, wrap the patient in a "bunny burrito" and release the rabbit on the floor to assess gait.	15a. This can be somewhat difficult, as a rabbit often will simply "freeze" because of the scent of other species that were in the examination room previously. A treat offered from the owner may tempt the rabbit to hop over to investigate. Rabbits hop forward using both forefeet together followed by the hind feet, front then back. They do not walk like other quadrapeds. Do not startle or "shoo" the rabbit to get it move. If it bolts, it may lose traction on the slick floor and injure itself.
16. Correctly restrain and lift the rabbit and put it back in its carrier as always—rear end first!	
17. Remove gloves and dispose of correctly. Hands of the examiner and the restrainer should be washed thoroughly in the examination room.	

UNIT SIX

Care of the Hospitalized Patient

OBJECTIVES

Upon completion of this unit, the reader should be able to:

▶ Provide appropriate housing and care for the hospitalized patient

GUIDELINES FOR CARE OF THE HOSPITALIZED PATIENT

When rabbits are hospitalized, they can easily become environmentally stressed, due in great part to visual and olfactory stimuli. Rabbits need to be housed in a quiet area away from the odors of predators, including dogs, cats, ferrets, snakes, and large birds. Rabbits never should be placed in cages facing predatory species where there may be visual contact, and this includes high-traffic treatment areas. One solution is to temporarily use a separate room entirely, be it a storage room or a clean isolation room with no other patients. Good ventilation and lighting are important.

The patient should be provided with a hide box or an area of the cage that is partially covered with a towel or blanket. Rationed pellets can be offered in a heavy ceramic bowl or a specially designed food hopper. Rabbits need to be provided with free-choice grass hay that is clean and dust-free, and a water sipper bottle attached to the front of the cage. Daily, they should also be given fresh greens and vegetables similar to those offered at home. The choices may vary, but there should never be an abrupt change in the diet. Whole carrots; carrot, turnip, and beet tops; and parsley are readily consumed. Lettuce, especially iceberg lettuce, should not be fed, as it has no nutritional value and can contribute to intestinal disorders. Although these foodstuffs should be offered, a rabbit that is ill may become anorexic and supplemental, force-feeding may be required.

Rabbits also need to be provided with a litter box, one dedicated for rabbits and not one normally used by cats. Instead of cat litter, the litter can be as simple as shredded newspaper or any other unscented substrate. Using a scented product may create avoidance, or depending on the type of substrate, may contribute to respiratory problems. Soiled bedding or fecal material initially placed in the litter tray will encourage the rabbit to use it.

PURPOSE

- To provide a quiet, yet readily accessible treatment cage

COMPLICATIONS

- None

EQUIPMENT

- Stainless steel cage
- Litter box
- Ceramic food bowl or hopper
- Sipper water bottle
- Towel, blanket, or hide box
- Supply of clean newspapers

PROCEDURE for Cage Set-up

TECHNICAL ACTION	RATIONALE/AMPLIFICATION
1. Line the cage floor with layers of newspaper.	1a. The cage should be no higher than chest height to care-givers. A rabbit should not have to be lifted out and down from a high cage because of the restraint concerns.
	1b. Always use non-colored sections of newspaper. When wet, many of the colors used in printing bleed, staining the patient's fur. It is also easier to observe urine or blood if it appears on non-colored newspapers.
2. Shred newspapers for litter box, or use another unscented type of litter.	
3. Put litter box in the back of the cage away from food and water.	
4. Put ¼–½ cup of rabbit pellets in bowl or hopper.	4a. If using a hopper, it should be attached to the cage door, near the water bottle.
5. Fill large sipper bottle with water, and attach to cage door.	
6. Put the hide box in the back at the opposite end of the litter box, or drape a clean towel so it covers half of the cage door.	6a. Suitably sized, clean cardboard boxes can be used, and they are easily disposed of after use. Do not use a modified plastic or rubber tub. Rabbits chew on most items, and both of these could be potentially dangerous if pieces are ingested or cause oral abrasions.
7. Place the patient in the readied cage, rear end first.	
8. Quietly close and latch the door.	8a. Do not startle the rabbit by quickly closing the door and slamming the latch into place. Always check to be sure that the latch is securely in place.
9. Observe the patient for a few moments.	9a. Depending on the health status of the patient, it may take a few tentative hops or "freeze." The patient assesses this new environment and either tries to bolt from it or accepts it as safe. If the cage is free from the odor of a potential predator, and there is no perceived external threat (visual or olfactory), the rabbit is likely to settle in quietly. If it seems overly anxious, cover the entire cage door and assess the environment for location, sights, smells, and sounds that may alarm the rabbit. After settling in and showing no signs of alarm or "thumping," the cage door can be partially uncovered.

Blood Collection

OBJECTIVES

Upon completion of this unit, the reader should be able to:

▶ Perform a blood draw from a rabbit

BLOOD COLLECTION

Blood collection sites in rabbits are similar to those in other small animals. The lateral saphenous or cephalic veins usually provide the easiest access and least amount of stress to the rabbit. The marginal ear vein or central artery of the ear is often cited for possible blood collection, but in reality it is difficult to obtain an adequate sample from either of these sites without causing trauma to the sensitive tissue of the ear, collapsing the vein, and creating a potentially large and painful hematoma. In a companion rabbit, ear draws should be avoided. The jugular vein can be accessed, but because of the restraint technique required, this procedure will cause a great deal of stress to the rabbit and potentially compromise an already critical patient.

Rabbits vary a great deal in weight, as does their blood volume. It is important to obtain an accurate weight at the time of the blood draw to calculate the amount that may be safely withdrawn from that rabbit at that specific time. As a *general* guideline, no more than 6% to 10% of the blood volume should be collected.

PURPOSE

- To obtain an adequate amount of blood to perform a complete blood count (CBC) and/or biochemistry panel

COMPLICATIONS

- Inadequate restraint technique
- Stress to the patient
- Hyperthermia
- Collapse of the vein
- Insufficient amount to perform required tests

EQUIPMENT

- 25–27 gauge needle
- 1 ml or tuberculin syringe
- Microtainer collection tubes (2) lithium heparin (green top), EDTA (purple top)
- Alcohol-soaked cotton balls
- Assistant for restraint
- Electric clippers with #10 blade attached

PROCEDURE for Blood Collection in a Rabbit

TECHNICAL ACTION	RATIONALE/AMPLIFICATION
1. Collect the necessary equipment.	
2. Examine available blood collection sites and decide which one is accessed more easily.	
3. Tell the restrainer which vein you will use so restrainer can position the patient.	
4. Shave a small area of fur over the site.	4a. Because of the density of rabbit fur, it is easier to visualize the vein if the area is shaved. Better visualization may prevent multiple puncture attempts. A #10 blade is sufficient; do not use #40 (surgical blade) at the risk of tearing the skin over the site.
5. Use alcohol-soaked cotton balls to clean the site.	
6. Hold the vein cranially to the puncture site and palpate the exact location.	6a. The person drawing the sample holds off the vein and completes the draw. This allows the restrainer's complete attention on the patient.
7. Insert needle with attached syringe and confirm with blood flash in hub of the needle.	
8. Draw only the amount required and no more than 6–8% of kilograms of body weight.	
9. Release pressure, and have the restrainer slide a thumb to the puncture site and apply firm pressure for 60 seconds or until all bleeding has stopped.	9a. 10a. Communicate with the restrainer, "withdrawing" (the needle) so he or she can slide a thumb over the puncture site and apply pressure. Rabbit blood clots quickly, so the sample has to be transferred to the appropriate collection tubes quickly. The tubes should be rocked gently to mix with the anticoagulants contained in the tubes, to avoid rendering the sample unusable.
10. Immediately transfer the blood sample to collection tubes, and gently rock the tubes.	

UNIT EIGHT

Medication Administration

OBJECTIVES

Upon completion of this unit, the reader should be able to:

▶ Administer medications; oral, subcutaneous (SQ), and intramuscular (IM)

GUIDELINES FOR ADMINISTERING MEDICATIONS

Rabbits will accept oral medications when given as a liquid or semi-liquid much better than they will accept a capsule or a tablet. Medication can be mixed with fruit juice, applesauce, mashed fruit (banana is a favorite), or pureed human infant food—fruit or vegetable. The mixture can be drawn into a syringe and administered orally if the rabbit refuses it. If supplemental feeding is required, rabbit pellets can be mixed with water or a flavoring, made soft and liquid enough to go through a syringe tip, or they can be placed in a blender with water or juice to achieve the correct consistency.

Administering IM injections causes pain and may elicit a scream from the rabbit. Always be prepared for this from an otherwise silent patient, as it can be startling and might cause the restrainer to let go of the patient. IM injections can also cause prolonged muscle inflammation and potential necrosis of muscle tissue. If there isn't an alternative route for the specific drug, the injection can be given in the quadriceps or into the dorsal (epaxial) muscles on either side of the spine.

Fluids are administered most often as a bolus subcutaneously, behind the scapula or along the back either side of the spine. The skin in these areas is loose, and the patient tolerates SQ administration of fluids without objecting too much. In a critically ill patient, an intravenous catheter can be placed. Sites for catheter placement include the cephalic or the lateral saphenous veins. The cephalic vein is less desirable because of the short foreleg. When using either site, the rabbit must be prevented from chewing through the line or pulling out the catheter. Placement of a catheter in the marginal ear vein is sometimes cited, but because of the sensitivity of the ear and the potential for permanent damage, using this site is not recommended.

A subcutaneous injection for a rabbit is straightforward. The injection can be administered most easily under the loose skin behind the shoulders.

PURPOSE

- To deliver the prescribed medications
- To provide supplemental feedings in an anorexic rabbit
- To deliver the prescribed medication with the correct frequency and volume

COMPLICATIONS

- Aspiration of an oral medication if the patient is restrained incorrectly
- Food/medication clogging syringe tip
- Intra-fur delivery of an injection instead of subcutaneously

EQUIPMENT

- 6–12 ml syringe for oral medications
- Prescribed medication
- Bowl
- Soaked or ground rabbit pellets
- Additions to pellets and medications to make them more palatable
- 22–25 gauge needle for injections
- 3 ml syringe for injections

PROCEDURE for Administering Oral Medication

TECHNICAL ACTION	RATIONALE/AMPLIFICATION
1. Have all equipment at hand and prepared.	
2. Draw up volume to be delivered into the syringe.	
3. Place the patient on table and hold the hind end tucked close to your body.	
4. Place the palm of your hand on top of the rabbit's head, and with the fingers, slightly retract the upper lips (**Figure 2-16**).	

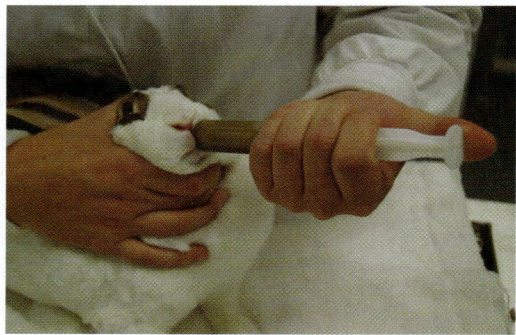

▲ **FIGURE 2-16** Oral medication or supplemental feedings can be administered to a rabbit via a syringe.

TECHNICAL ACTION	RATIONALE/AMPLIFICATION
5. Insert the syringe tip from the side of the mouth just behind the incisors.	5a. The oral cavity of a rabbit is small, and the mouth does not open wide. Putting the syringe tip into the side of the mouth provides enough room without discomfort to the patient. Do not attempt to force the mouth open.
6. Administer in small increments, allowing the rabbit to taste and swallow normally.	6a. Make sure that the food is not placed in the cheek or the rabbit will not swallow it normally.
	6b. Always keep the head parallel to the surface to avoid aspiration. Never attempt to administer anything orally by holding the rabbit upside down, cradled in your arms.
	6c. Rabbits that refuse to swallow or grind their teeth may be suffering from abdominal pain.

PROCEDURE for Administering a Subcutaneous Injection

TECHNICAL ACTION	RATIONALE/AMPLIFICATION
1. Push the patient to the corner of the cage.	1a. Usually, no further restraint is required if the rabbit is positioned so its rear end is in a corner of the cage.
2. Gather scruff.	
3. Inject subcutaneously just behind the shoulders under the skin below the scruff.	

PROCEDURE for Administering an Intramuscular Injection

TECHNICAL ACTION	RATIONALE/AMPLIFICATION
1. Secure the patient in a towel wrap ("bunny burrito"). If using the quadriceps, leave one hind leg unwrapped.	1a. A secure restraint is required for an IM injection because of the pain it causes to the patient. In addition to the injection itself, many medications sting or burn.
2. Restrainer must have a secure hold and not release the patient.	
3. Locate injection site and administer.	
4. Gently massage the muscle.	4a. Massaging helps disperse the medication and decrease the pain.
5. Return the patient to the cage, placing rear end in first, then release the rabbit from the towel.	5a. Take great care in releasing the patient, as it is likely to attempt to bolt from the perceived source of pain.

UNIT NINE

Anesthesia

OBJECTIVES

Upon completion of this unit, the reader should be able to:

▶ Induce and maintain anesthesia

GUIDELINES FOR RABBIT ANESTHESIA

Rabbit anesthesia can be a little more of a challenge than with other small mammals. Being aware of some of the idiosyncrasies can make induction and recovery less stressful for the rabbit. Rabbits do not need to be fasted prior to anesthesia and they are unable to vomit due to the anatomical placement of the stomach. Many patients presented for surgery or a procedure requiring anesthesia may have been anorexic but it is always important to discuss food consumption with the owner to determine what, if anything was given or consumed.

The inhalant anesthetic agents used in rabbits are either isoflurane or sevoflurane. Both are well tolerated in rabbits, but an advantage of sevoflurane is that it doesn't have an odor, making it easier to mask down the patient. Rabbits often hold their breath when being induced, and resist breathing in the anesthetic agent. This is especially evident when rapid induction is practiced. For the rabbit, a slow and calculated induction is preferable. Before applying the face mask, allow a few minutes of pure oxygen to flow through the system and mask, which will help remove any residual odor in the mask.

Intubation of a rabbit is difficult because the oral cavity is so small. The tongue is thick and further blocks visualization of the epiglottis, which is more caudal than in other species. These anatomical differences all contribute to the difficulty of intubation.

A WORD OF WARNING

Never assume that a rabbit has reached a suitable plane of anesthesia by visual parameters alone. Many a rabbit has been placed in dorsal recumbency, shaved and scrubbed for the surgical table, apparently anesthetized, only to bolt off the table in an attempt to escape.

PURPOSE

- To induce and maintain anesthesia for a surgical procedure
- To use anesthesia as a method of restraint for radiographs
- To alleviate pain during a clinical procedure

COMPLICATIONS

- Willful apnea (resistance to breathe anesthetic gas)
- Difficulty placing an endotracheal (ET) tube
- Misjudging anesthesia plane

EQUIPMENT

- Small-animal face mask
- Standard small-animal anesthesia machine with non-re-breathing delivery system
- Eye lubricant
- If attempting endotracheal intubation; 2.0–4.0 mm uncuffed endotracheal tube
- Local anesthetic (0.2% lidocaine)
- Narrow-blade laryngoscope

PROCEDURE for Inducing and Maintaining Anesthesia Using a Face Mask

TECHNICAL ACTION	RATIONALE/AMPLIFICATION
1. Place the patient on the table and hold it with the rear end tucked close to your body.	
2. Apply eye lubricant.	
3. Turn the flow of oxygen to 100% (maintenance will be to 100–300 ml/kg/minute).	
4. Allow 100% oxygen to flow through system and face mask for 2–3 minutes.	
5. Gently position the face mask around the nose and mouth to secure a correct fit with no ocular pressure.	
6. Allow the rabbit to breathe 100% oxygen for another few minutes before introducing the anesthetic agent.	
7. Introduce gas slowly in 0.5% increments, and adjust oxygen to maintenance level.	7a. When anesthesia gas is introduced in 0.5% increments, the occurrence of apnea (breath holding) decreases, making for a smoother induction.
8. Monitor patient carefully for respiration rate.	

PROCEDURE for Endotracheal Intubation

TECHNICAL ACTION	RATIONALE/AMPLIFICATION
1. Follow anesthesia procedure steps 1–8, above, to induce the patient.	
2. With the use of a 1 ml syringe (needle removed), place 2 or 3 drops of 0.2% lidocaine in the area of the larynx.	2a. The rabbit has a long, narrow oral cavity and the oropharyngeal area is very sensitive. Using a few drops of a local anesthetic can help reduce laryngospasms.
3. The restrainer holds the head so the head is pointed up, perpendicular to the body, retracts the front lips, and holds the upper jaw (**Figure 2-17**).	

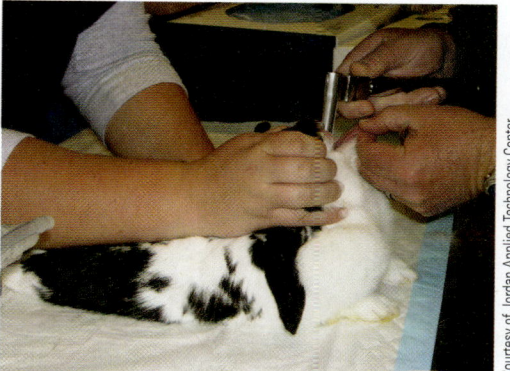

Courtesy of Jordan Applied Technology Center

▲ **FIGURE 2-17** The restraint position required in an anesthetized rabbit is important for successful placement of the ET tube; a laryngoscope is necessary to visualize the epiglottis.

(Continues)

TECHNICAL ACTION	RATIONALE/AMPLIFICATION
4. Gently pull the tongue out and slightly down.	
5. Introduce laryngoscope.	
6. Introduce ET tube, guiding it by the laryngoscope into the epiglottis (the opening to the trachea).	6a. Do not make repeated attempts to intubate the patient. Stimulation during the attempts can disrupt the plane of anesthesia achieved and further increase the frustration for all involved and potentially cause trauma to the rabbit. Intubation of the rabbit needs patience and practice.
7. Slowly advance ET tube each time the rabbit takes a breath, until successfully positioned within the trachea.	
8. When placement is confirmed, remove the laryngoscope and attach the patient to the anesthesia delivery system.	

Surgical Preparation

OBJECTIVES

Upon completion of this unit, the reader should be able to:

▶ Prepare the patient for a surgical procedure

KEY TERMS

atropinase

GUIDELINES FOR SURGICAL PREPARATION OF THE RABBIT

Preparation for surgery is not a great deal different from that of other small animals. One thing to note is that some rabbits produce **atropinase**, an enzyme that renders atropine sulfate ineffective. There is no way to predetermine the effect that atropine might have on a given rabbit. Glycopyrrolate provides the same effect as atropine sulfate and it is well tolerated and effective in rabbits. Always check with the veterinarian regarding pre-op protocols that he or she prefers to use in the rabbit patient.

PURPOSE

- To prepare the patient for abdominal surgery
- To prepare the patient for a localized area of a surgical procedure

COMPLICATIONS

- Inducing a trance-like state instead of general anesthesia

EQUIPMENT

- Electric clippers with #40 blade attached
- Scrub soap and solution
- Alcohol or providone-iodine solution (Betadine)
- Disposable gloves

PROCEDURE for Surgical Prep of the Rabbit

TECHNICAL ACTION	RATIONALE/AMPLIFICATION
1. Follow steps 1–8, above, for inducing anesthesia with the use of a face mask.	
2. When the patient has reached a suitable anesthetic depth, lay the rabbit in dorsal recumbency (or in position to prepare a non-abdominal incision, depending on the surgical site).	
3. Using the #40 blade on the clippers, shave the area cleanly, use vacuum to collect all loose fur.	
4. Put on disposable gloves and proceed with the standard aseptic technique for a surgical scrub.	
5. Use Betadine solution or alcohol for a final rinse of the incision site.	
6. Carefully and continually monitor anesthetic depth.	6a. 7a. Trance-like state or anesthesia? Check the following for assessing anesthetic depth: The nictitating membrane moves approximately 1/3 across the cornea, and there is a loss of palpebral reflex, there is no reaction to a toe or *gentle* ear pinch, and the abdominal muscles are relaxed. Respiration decreases to 18–24 b/min. These indicators occur only when the patient is anesthetized.

(Continues)

TECHNICAL ACTION	RATIONALE/AMPLIFICATION

7. Do not turn your back on the patient or leave the patient for any reason.

8. Transfer patient to the surgical area or the OR.

9. Table restraints (ties), positioning, and monitoring devices are the same as for a cat or a small dog (**Figure 2-18**).

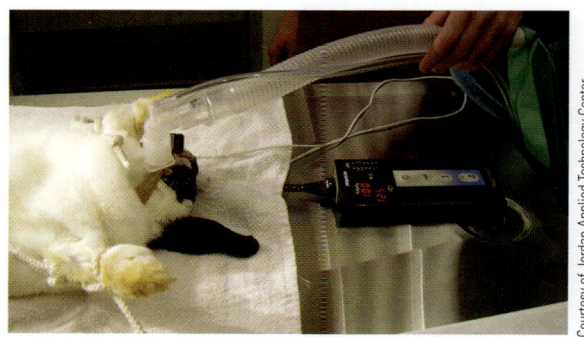

Courtesy of Jordan Applied Technology Center

▲ **FIGURE 2-18** A rabbit positioned for surgery and connected to a non-re-breathing anesthesia system.

Radiology

OBJECTIVES

Upon completion of this unit, the reader should be able to:

▶ Take standard view radiographs; ventral/dorsal and lateral

GUIDELINES FOR RADIOLOGY TECHNIQUES OF THE RABBIT

Radiograph techniques in the rabbit are similar to those used with other small mammals. Settings for the X-ray machine approximate those used with a cat. Frequently, views requested are ventral/dorsal and lateral of the abdomen (**Figure 2-19**). Abdominal radiographs can diagnose gut-stasis with multiple air pockets (Figure 2-19). Frequently, radiographs of the head can demonstrate the presence of abscesses or dental problems. Extremities are radiographed to determine the exact location and severity of a fracture (**Figure 2-20**).

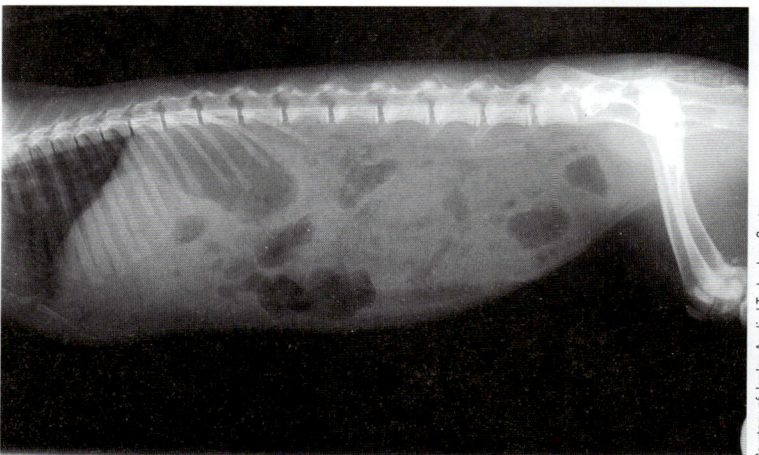

Courtesy of Jordan Applied Technology Center

▲ **FIGURE 2-19** A lateral radiograph of a rabbit showing multiple air pockets throughout the intestinal tract.

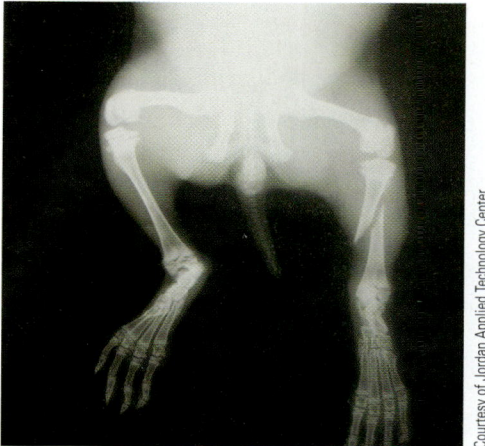

Courtesy of Jordan Applied Technology Center

▲ **FIGURE 2-20** This radiograph clearly defines the fracture of the left tibia. The rabbit was accidentally caught under the rocker of a chair.

Patients should be masked down with light anesthesia to avoid causing pain or further trauma during manipulation and positioning.

PROCEDURE for Radiographs in the Rabbit

TECHNICAL ACTION	RATIONALE/AMPLIFICATION
1. Have 2 film cassettes or digital imaging plates ready for use.	
2. Follow steps 1–8 of administration of anesthesia to mask down the rabbit.	
3. Measure area to be radiographed, and adjust settings on the X-ray machine.	3a. The technique chart used for the cat is approximate. Always obtain your own measurements to avoid having to take multiple exposures.
4. Position the patient: If lateral, the body should be extended. If V/D, the body should be straight with the hind legs extended caudally.	
5. Take radiographs, and recover the patient.	

FAST FACTS

Weight

- **Buck adult weight:** 2–5 kg (4.4–11 pounds) average
- **Doe adult weight:** 2–6 kg (4.4–13.2 pounds) average

Life Span

- 7–9 years

Reproduction

- **Sexual Maturity**
  - **Bucks:** 22–25 weeks
  - **Does:** 22–25 weeks
  - **Gestation:** 30–32 days
  - **Litter size:** 4–8 kits
  - **Weaning age:** 4–6 weeks

Vaccinations

- none

Vital Statistics

- **Temperature:** 38.3°–40° C (101°–104° F)
- **Heart rate:** 130–325 breaths/minute
- **Respiratory rate:** 35–60 breaths/minute

Dental

- Dental formula 2 (2/1,0/0,3/2,3/3) = 28 total teeth

Zoonotic Potential

➤ Fungal: Ringworm (Dermatophytosis)
➤ Bacterial: Salmonella, Tularemia, Leptospirosis, Streptobacillus, Pasteurella

Review Questions

1. In what circumstances would a rabbit vocalize?

2. Why is correct restraint of the rabbit so important?

3. What are cecotrophs, and why are they important to a rabbit's health?

4. What potential problem could arise if a rabbit is continually fed a high-protein diet?

5. What are papillomas?

6. What could be the consequences of attempting to squeeze a lump or bump under the skin?

7. Why is it sometimes necessary to trim the incisor teeth of a rabbit?

8. What is the significance of a head tilt in a rabbit?

9. List the advantages and disadvantages of inducing a trance-like state.

10. Why is intubation more difficult in a rabbit than in other species?

11. What is atropinase, and how does it affect pre-surgical protocol with rabbits?

12. List five indicators of anesthetic depth in a rabbit.

13. What is the average gestation period in rabbits?

14. Which zoonotic disease can be transmitted by direct contact?

15. What are the normal ranges for respiration, heart rate, and temperature?

References

Crow, Steven E., & Walshaw, Sally O. *Manual of Clinical Procedures in the Dog, Cat and Rabbit, 2nd ed.* (1997). Ames, IA: Lippincott Williams and Wilkins.

Judah, V., & Nuttall, K. *Exotic Animal Care & Management, 2nd ed.* (2014). Clifton Park, NY: Cengage Learning.

Warren, D. *Small Animal Care & Management, 3rd ed.* (2010). Clifton Park, NY: Cengage Learning.

http://www.floridarabbit.org (accessed July 31, 2011)

http://www.rabbit.org (accessed November 16, 2011)

http://www.bunnycentral.com (accessed July 30, 2012)

http://www.bu.edu (accessed June 17, 2014)

CHAPTER 3

FERRETS

A ferret is God's way of telling you
that nothing is childproof.

—*Unknown*

Overview of Species

OBJECTIVES

Upon completion of this unit, the reader should be able to:

▶ Provide appropriate client education to new ferret owners

KEY TERMS

congenitally	hypoglycemia
free choice	jills
gib	kits
hobs	sprite

FERRETS

▲ **FIGURE 3-1** Ferrets are playful and amusing pets.

The domestic ferret is a delightful, fun-loving companion animal (**Figure 3-1**). It belongs to the family *Mustelaide*, which also includes skunks, weasels, mink, badgers, and otters. Ferrets have been domesticated for centuries and originally were kept as hunting animals and for pest control. They are efficient hunters with the strong instincts of a predator.

Ferrets sold in pet shops most often come from commercial breeding farms. Commercial ferret breeders spay, neuter, and descent the **kits** (babies) around 5 weeks of age and give them an initial vaccination for canine distemper. Ferrets that are bred by one large farm—Marshall Farms in New York, for example— have two tattooed dots in their right ear designating that these procedures have been performed.

Ferrets may not be legal in some areas, so staff members should be aware of not only state laws but also local ordinances disallowing ferrets. Even though a state law may not ban ferrets, local and municipal laws may prohibit them. Also, in the United States, it is illegal to hunt with a ferret.

There are now many color variations in ferrets. They vary from their normal sable coloring to albino, butterscotch, silver mitt, and pied (a base color with white patches in the coat). Some of the newer color variations have distinctive white markings on the head and body. Because of the inbreeding required to establish new coat colors, many ferrets with white markings on their heads may be **congenitally** deaf—born without the ability to hear.

Ferrets of either sex make equally good pets. **Hobs** (males) can be twice the size of **jills** (females). This is especially evident if hobs were not neutered at an early age. Hobs weigh an average of 2–4.5 pounds (1–2 kg) Jills often weigh as little as 1–2.5 pounds (0.6–0.9 kg) A neutered male is referred to as a **gib**. A spayed female is called a **sprite**. Both sexes have a musky odor even when they have been descented. They have small scent-producing sebaceous glands on the abdomen, which cannot be removed surgically. By keeping the bedding and litter box clean, the odor can be reduced greatly.

BEHAVIOR

Ferrets are playful and often engage in games with their owners. An excited, playful ferret performs an array of movements by jumping hunch-backed, bouncing up and down, running backward or around in tight circles, shaking its head and making soft chuckling sounds and hissing.

Play behavior between ferrets is rough, especially with kits. Like kittens that learn to stalk, pounce, and bite, young ferrets have no hesitation in being just as rough with people. Kits grab and bite each other's scruff, or drag each other by the ears or any other appendage. They should be taught at a young age that biting and pulling are not so enjoyable for humans. Ferrets are intelligent and responsive to the human voice, quite capable of understanding the word "No!" Scruffing a ferret by the area of skin at the nape of the neck and saying "No" in a firm voice is effective. A simple, painless scruff is all that is required—a "time out" so to speak.

Ferrets should never receive physical punishment. One of the worst (and unfortunately common) recommendations to stop a ferret from biting is to flick the ferret on the end of the nose. This is not only painful for the ferret and completely unjustified, but it teaches the ferret that hands cause pain, and the lesson for the ferret is one of avoidance or aggression.

Ferrets are also notorious thieves. They will take anything that appeals to them and stash the item away. Missing items are usually found all together in one place. Once owners discover the stash, they are likely to find things they didn't even know were missing: car keys, wrapped sweets, a single sock, prescription bottles, cash, or wallets. Clearly, these are not food items for the ferret, and the reason for this thieving/hoarding behavior is unclear. Wise owners know probable locations of stashes and have learned to look there first for any missing items. A good place to begin the "treasure hunt" is the inside of a box-spring mattress and in the back of the family sofa or an upholstered chair.

Although ferrets enjoy playing with toys, they must be chosen carefully. Ferrets often chew off small pieces and swallow them, consequently requiring surgery for a foreign-body obstruction. The biggest danger is not from toys per se but, rather, other items found around the household. Common causes of intestinal blockage are small pieces of foam rubber, pencil erasers, rubber bands, wads of thread, and Velcro®. (Shoes with this type of fasteners and rubber soles should be examined carefully for missing pieces.)

The spine of a ferret is long and flexible, allowing ferrets to maneuver through small spaces and still be able to turn around. Ferrets will investigate any space and are capable of flattening their bodies to squeeze into and under small areas. For this reason, owners must "ferret-proof" their homes, as a ferret can disappear into a small space that the owner may not have noticed.

Ferrets are active diggers and can throw out all of the soil from a potted plant in a matter of minutes. They will dig at carpets and in the corners of rooms. Soft furnishings are a favorite play area; a ferret will go in, over, and under, pulling itself along on its back, potentially causing as much damage as a cat clawing at the furniture.

✳ DECLAWING

Occasionally, frustrated owners may ask about declawing their ferrets. Declawing a cat is controversial, but declawing a ferret is crippling and the ferret will suffer a permanent disability. The claws of a ferret are not retractable, and the anatomy of the digit is completely different. "Declawing" would include not only removing the claw but also amputating the last joint of each toe, making it difficult for the ferret to walk correctly.

Ferrets spend a great deal of their time sleeping, and it is not unusual for a healthy ferret to sleep as much as 18–20 hours in the 24-hour day. The sleep is very deep, especially in kits, and many new owners have become alarmed, thinking that their pet might be in a coma or even dead.

Because one of the things ferrets do best is to sleep, elaborate cages are not required. Many ferrets are housed in sturdy pet carriers. Ferrets do not play in their sleeping dens. They need only a water bottle, a full cup of food, a pile of soft bedding (old towels, small blankets, a discarded sweatshirt), and a litter box at the opposite end of the carrier.

Ferrets will use a litter box provided that it is in the right place at the right time. During play, they will not actively seek out a litter box but instead will run backward to the nearest corner of the room. Ferrets elevate the rear end to urinate and defecate, so protected corners of rooms should have a small area of the wall covered. Litter boxes specifically designed for ferrets are triangular (to fit into corners), and the back of the triangle is higher to accommodate the ferret's elimination behavior. Basic, unscented cat litter is the best choice, as clumping litters have the potential for creating an intestinal blockage if ingested.

DIET

As carnivores, ferrets require a diet based on animal-derived protein which can be found in a variety of diets specifically formulated for ferrets. Many ferrets are fed, and will eat, a variety of fruits and dairy products, but these contribute to digestive and urinary tract problems. Abrupt or random changes in brands should be avoided to prevent diarrhea and intestinal upset.

Ferrets have a short gastrointestinal tract and transit time. Food should be available **free choice**, allowing the ferret access to food at all times. They do not nibble at food but, instead, eat from the bowl until their needs are met. Restricting food may cause **hypoglycemia**, a condition of low blood sugar.

With a high-quality diet, nutritional supplements are not needed, but vitamin and mineral supplements are available that the ferret will take readily. These products can improve skin and coat condition and assist in preventing the build-up of hair in the intestinal tract and the formation of hairballs. When used as a treat, most ferrets will abandon any activity or hiding place for the proffered bribe.

HEALTH CARE

Ferrets are highly susceptible to canine distemper and do not survive the disease. They should be vaccinated with a distemper vaccine approved for use in ferrets, and under no circumstances should a canine distemper "combo" vaccine be used. Vaccination against rabies is recommended and, in some areas, required. (Refer to the vaccination schedule in Fast Facts). Ferrets are also susceptible to the human influenza virus, so owners should be made aware that their pet can catch "the flu" from them. With supportive care and treatment, ferrets often survive what can be a severe respiratory disease, the same as in humans.

BREEDING

Sexing of ferrets is similar to that of dogs. The penis is mid-abdomen, caudal to the umbilicus. The vulva of the immature female is small but clearly discernable below the anus. Ferrets are seasonal breeders; their breeding cycle is controlled by the number of daylight hours. The jill is an induced ovulator and will remain receptive (in heat) unless bred. The increased levels of estrogen will lead to estrogen toxicity and, in turn, cause bone marrow suppression and severe anemia. Although it is possible to stop the breeding cycle with hormones, recommended treatment is to either breed the jill or spay her. The gestation period is 42–45 days. The young are born altrical, and the average litter size varies from 6–8 kits.

Restraint Techniques

OBJECTIVES

Upon completion of this unit, the reader should be able to:

▶ Demonstrate correct methods of restraint for a variety of procedures

GUIDELINES FOR RESTRAINT AND HANDLING OF THE FERRET

Ferrets can be picked up easily with one hand, just behind the shoulders. With all four feet off the surface, they usually relax and may be carried in this manner without a struggle. For other than removing from or returning a ferret to its cage, the hind end should be supported (**Figure 3-2**). Ferrets should not be expected to ride around on the shoulders or in the back of a hooded sweatshirt or handled in any way that would subject them to a fall.

▲ **FIGURE 3-2** The correct technique for holding a ferret. The rear quarters should be supported.

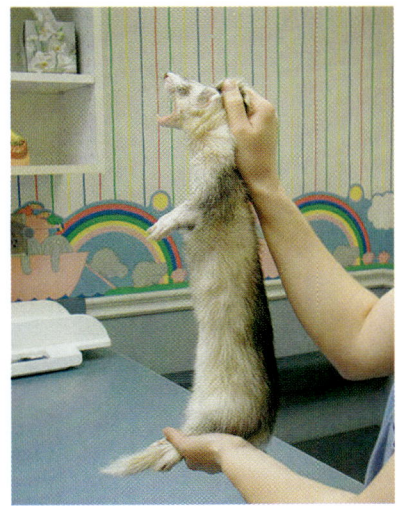

▲ **FIGURE 3-3** When first scruffed, a ferret will normally elicit a wide yawn, and this provides an opportunity for a quick examination of the oral cavity.

Restraint for an exam can be more difficult because of the ferrets' flexibility and quickness. Scruffing the ferret is safe and effective in all except ferrets that are frightened, injured, or in pain. Many ferrets relax so completely with a maintained scruff that they can be laid down and positioned for a radiograph without moving and without the use of general anesthesia. This is also a good technique for a quick exam of the oral cavity. Ferrets have a yawn reflex associated with being scruffed, and the ferret will open its mouth wide as it yawns (**Figure 3-3**).

PURPOSE

- To restrain a ferret for a variety of procedures including nail trims, ear cleaning, taking a rectal temperature, giving injections, and positioning for a radiograph.

COMPLICATIONS

- The patient breaks free from the scuff and falls

EQUIPMENT

- None

PROCEDURE for Restraining a Ferret with a Scuff

TECHNICAL ACTION	RATIONALE/AMPLIFICATION
1. Pick up the ferret with one hand.	
2. With the other hand, firmly grasp the area of skin behind the neck.	2a. In large hobs the scruff may be a little harder to grasp, as the skin is tight and very tough.
3. Lift the ferret by the scruff so all four feet are off the surface.	3a. Do not hold the patient directly over the floor. If the ferret breaks free from the scruff and falls, serious injury could result.
4. Place the free hand around the rear limbs, just below the pelvis.	4a. If the ferret starts to twist and squirm, a two-handed hold provides for greater control.

Grooming

OBJECTIVES

Upon completion of this unit, the reader should be able to:

▶ Perform grooming: nail trims, bathing, and ear cleaning

KEY TERMS

exudate

GUIDELINES FOR GROOMING THE FERRET

Ferrets may be given a bath with a shampoo specifically formulated for ferrets or one made for kittens. Ferrets should not be bathed more frequently than once every three months. Frequent bathing actually increases sebaceous gland secretion in compensation for the oils lost.

The nails should be trimmed regularly. Each foot has five claws, which can grow long, especially the front claws, and can become entangled in carpet and bedding.

All ferrets produce a grainy, reddish-brown **exudate** from their ears. This is often mistaken for, or assumed to be, an infestation of ear mites. If the debris is very dark and has an odor, ear mites can be suspected. An examination of the material under the microscope can confirm if ear mites are present, as well as the normal exudate.

Many clients request that all three grooming tasks be performed in one visit. It works well to perform them in this sequence: nail trim, ear cleaning, and bath.

PURPOSE

- Trim nails to remove overgrowth
- Clean ears and remove exudate and evaluate for presences of ear mites.
- Bathe a ferret to help remove odor, loose hair, and flaky skin

COMPLICATIONS

- Cutting into the nail quick
- Penetrating the ear canal with a cotton-tipped applicator
- Getting shampoo into the ferret's eyes

EQUIPMENT

- Small cat claw trimmers or human nail clippers
- Cat hairball remedy
- Tongue depressor
- Assistant for restraint of feisty patient
- Cotton-tipped applicators
- Ear-cleaning solution
- Ferret or kitten shampoo
- Terrycloth towels

PROCEDURE for a Ferret Nail Trim

TECHNICAL ACTION	RATIONALE/AMPLIFICATION
1. Scruff the ferret and hold it above padded exam table with one hand.	1a. Never hold a ferret directly above the floor in case it breaks away from the scruff and falls. Use folded towels for the padding on the exam table and then for towel-drying after the bath. Towel-dry ferrets instead of putting them into a cage with a blow drier, which can be too hot, and the ferret may escape through the cage bars.

(Continues)

TECHNICAL ACTION	RATIONALE/AMPLIFICATION

2. Using either cat claw trimmers or human nail clippers, cut each claw just short of the quick (**Figure 3-4**).

2a. If using human nail clippers, turn the cutting edge perpendicular to the nail to prevent splitting the nail.

◀ **FIGURE 3-4** Clipping the ferret's nails with a simple scruff restraint technique.

3. If the patient is reluctant to cooperate, engage an assistant for restraint.

4. Have the restrainer squeeze a small amount of the cat hairball remedy onto the tongue depressor and allow the ferret to lick it while the other person performs the trim.

4a. Use the "scruff and dangle" approach with another person maintaining the scruff, to make the task quicker and easier and reduce the chances for inadvertently cutting into the quick or cutting a toe.

4b. Distract ferrets by offering a cat hairball remedy on a tongue depressor. Ferrets are fond of the malt flavor, and they will be distracted enough not to object to the nail trim.

PROCEDURE for Ear Cleaning

TECHNICAL ACTION	RATIONALE/AMPLIFICATION

1. Scruff the ferret in one hand.

2. With the other hand, place a few drops of ear-cleaning solution in each ear and massage.

3. Use cotton-tipped applicators to clean out debris, being careful not to enter the ear canal (**Figure 3-5**).

4. Continue to clean until the ear is free of all debris.

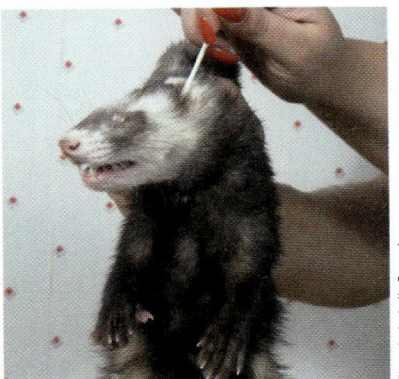

▲ **FIGURE 3-5** Cleaning a ferret's ears is easily accomplished by using the same scruff technique.

(Continues)

TECHNICAL ACTION	RATIONALE/AMPLIFICATION
5. If exudate is very dark and has an odor, set a dirty cotton tip aside for examination under the microscope.	5a. Label the microscope slide with the patient's number. Roll the cotton-tipped applicator across the slide, add one drop of water or immersion oil, place cover slip over the slide, and examine at 40 × and then 100 ×. *Otodectes cynotis*, the ear mite found in dogs and cats, also infects ferrets.

PROCEDURE for Bathing a Ferret

TECHNICAL ACTION	RATIONALE/AMPLIFICATION
1. Hold the ferret in one hand, and do not immerse in the water.	1a. Ferrets are not fond of water but usually will not resist when held with all four feet out of the water (**Figure 3-6**).

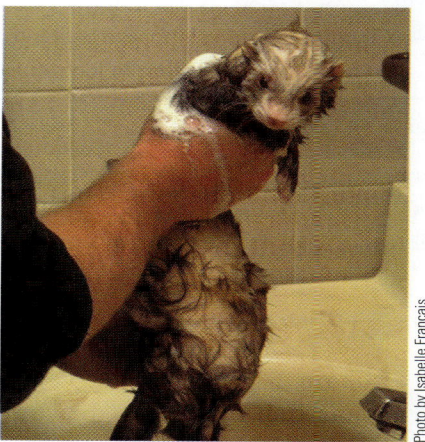

Photo by Isabelle Francais

▲ **FIGURE 3-6** Ferrets do not usually resist being bathed.

TECHNICAL ACTION	RATIONALE/AMPLIFICATION
2. Adjust the water flow, and monitor water temperature carefully.	2a. The water flow should be gentle, not a blasting spray, and the temperature should feel slightly warm, but never hot.
3. Thoroughly wet the ferret's coat.	3a. Take care not to get any shampoo in the eyes. Using a sterile eye lubricant first will help protect the eyes from soap.
4. Apply shampoo and rub in well, adding more water to the coat as necessary.	
5. Rinse the coat completely.	
6. Towel-dry and place the ferret in a warm place to finish drying.	

Patient History

OBJECTIVES

Upon completion of this unit, the reader should be able to:

▶ Complete patient history

GUIDELINES FOR TAKING A PATIENT HISTORY

The patient history is completed on the medical record form. In particular, pay close attention to how the client describes the reason for the visit or the chief complaint. Record exactly how the client describes a condition, concern, or change in behavior. Listen and carefully record what the client says, which often provides a great deal of information for the veterinarian.

PURPOSE
- Record patient signalment: age, sex, and reproductive status
- Determine the client's chief complaint, the reason for the visit
- Obtain information on diet and husbandry
- Record vaccination status
- Determine if anyone in the household has recently been ill with influenza

COMPLICATIONS
- Incomplete history
- Poor communication with client

EQUIPMENT
- Medical record form
- Pen (if not entered directly into the patient's computer file)

PROCEDURE for Taking a Medical History

TECHNICAL ACTION	RATIONALE/AMPLIFICATION
1. Enter objective information.	
2. Ask about diet, housing, bedding, and treats given.	
3. Obtain detailed information regarding the reason for the visit.	
4. Determine how long signs have been present.	
5. Ask if the client has treated the patient with anything and, if so, the result.	
6. Ask for the vaccination status of the patient: when the vaccine was given, what was given, how it was given, and if there were any reactions to the vaccines.	6a. If the ferret was not vaccinated by a veterinarian, it is important to establish the specific vaccine used and where it was obtained. With the exception of rabies vaccine, pet stores sometimes offer vaccination services, or in some instances vaccinations can be bought over the counter from feed stores and large chain grocery store pharmacies. If possible, obtain the specific type of vaccine, where it was given, by what route, and see if it can be determined that the vaccine was handled and stored correctly prior to use (was it refrigerated, pre-mixed, drawn into a syringe from a multidose vial?).

(Continues)

TECHNICAL ACTION	RATIONALE/AMPLIFICATION
	6b. Ask how long ago the vaccination was administered.
7. Ask if anyone in the household has recently been ill with the flu.	**7a.** Ferrets are susceptible to the human flu virus. If a client reports that someone had or has a severe "cold," that should also be noted, as it may have been an undiagnosed case of influenza.

UNIT FIVE

Physical Examination

OBJECTIVES

Upon completion of this unit, the reader should be able to:

▶ Perform basic health examination

KEY TERMS

Alopecia

GUIDELINES FOR A PHYSICAL EXAMINATION

Physical examination of the ferret follows the same procedure as for a dog or a cat. Visually examine all systems, and record objective findings. If there is an area of concern either during the examination or expressed by the client (other than the chief complaint), note that in the examination form.

PURPOSE

- To complete the basic details of the patient's status with regard to observable signs and conditions
- To obtain weight
- To obtain temperature, pulse, and respiration
- To record findings for the veterinarian

COMPLICATIONS

- None

EQUIPMENT

- Pediatric stethoscope
- Scale
- Thermometer
- Lubricating gel
- Paper towels
- Tongue depressor
- Tube of cat hairball remedy
- Assistant for restraint when obtaining rectal temperature

PERFORMING the Physical Examination

TECHNICAL ACTION	RATIONALE/AMPLIFICATION
1. Place a towel on the examination table. Stay close!	1a. Examination tables can be slippery, and ferrets are very nearsighted. Allowing the ferret to "explore" could lead to a fall, causing a potentially serious injury.
2. Record overall appearance of the patient with regard to coat, areas of hair loss, and body mass.	2a. If there are specific areas of **alopecia,** describe the location and ask the client how long the hair loss has been evident.
	2b. Alopecia is one of the first signs of adrenal gland disease. Hair loss starts on the rump, then progresses across the hips and down the tail. In the sprite, the vulva becomes swollen and it may appear as if she is in heat. Male ferrets with adrenal gland disease have trouble urinating. Both sexes may exhibit edema (**Figure 3-7**).

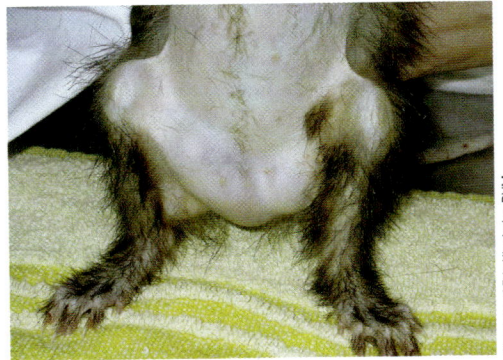

Courtesy of Eric Klaphake, DVM

▲ **FIGURE 3-7** A typical sign of hair loss and edema seen frequently associated with adrenal gland disease.

(Continues)

TECHNICAL ACTION	RATIONALE/AMPLIFICATION

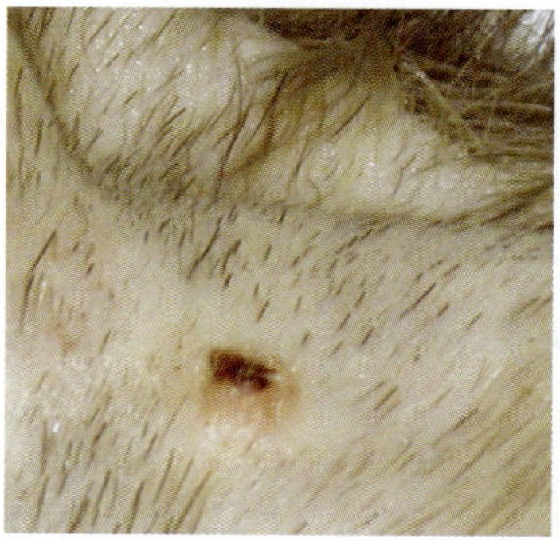

Courtesy of Eric Klaphake, DVM

▲ **FIGURE 3-8** A mast cell tumor may initially resemble an insect bite.

2c. Mast cell tumors are often diagnosed in the ferret. When examining the skin, look for small, hairless areas that may appear to be an insect bite or scratch. In later stages, they may be swollen, black, and crusty. It is important that these small tumors are diagnosed early (**Figure 3-8**).

3. Hold the ferret with one hand, and examine the eyes and ears.

3a. The eyes should be clear and free from any discharge.

3b. More likely than not, the ears will be dirty unless the owner cleans them regularly. The veterinarian may recommend an ear-cleaning and maybe a microscopic look at the debris to see if ear mites are present. Part of the technician's follow-up to the veterinarian's exam may be to show the client how to clean the ferret's ears.

4. Scruff the patient. When the ferret yawns in response, quickly look in the oral cavity.

4a. When the ferret yawns, look at all four canines. It is not uncommon for a ferret to have a broken canine. Further examination of the oral cavity might require anesthesia if a dental problem is suspected.

5. Place the patient on the table, holding with one hand behind the shoulders and around the abdomen.

6. Obtain the heart rate with the stethoscope.

7. Observe and record the respiration rate.

8. Lubricate the thermometer.

8a. With practice, one person can obtain a rectal temperature using just a scruff. Hold the scruff with one hand, lift the ferret so all four feet are away from the surface. When the thermometer tip is inserted, wrap your index finger around the tail and the thermometer to hold it in place. Some otherwise calm ferrets object mightily to this, and it is better to ask for an assistant to restrain the ferret rather than to have it struggle. The assistant can "scruff and dangle" and offer the patient a cat hairball remedy on the end of a tongue depressor. This usually provides enough distraction for the patient.

9. Re-scruff the ferret and gently insert the thermometer tip into rectum. Record the body temperature.

10. Return the ferret to the carrier or the client, ready for consultation with the veterinarian.

Care of the Hospitalized Patient

OBJECTIVES

Upon completion of this unit, the reader should be able to:

▶ Care for the hospitalized patient

© Eric Isselee/Shutterstock.com

GUIDELINES FOR CARE OF THE HOSPITALIZED FERRET

Hospitalized ferrets can be easily accommodated in small animal crates, such as a hard dog or cat carrier. If they are housed in standard kennel cages, there is a risk of injury if they escape through the bars, fall to the floor, and disappear somewhere in the clinic or hospital, as they are inclined to do. A loose ferret may also put other small animals and avian patients at risk.

Ferrets need to be provided with a sipper bottle of fresh water like those used for rabbits and rodents, and they must have access to food at all times. They should have ample soft bedding. Old towels and blankets are all a ferret needs for comfort and security. The bedding should be changed daily, or more frequently if soiled. Items used should be easily laundered without the addition of fabric softeners or drying sheets. The carrier should also be large enough to put a tray of plain litter in a back corner.

Supportive feeding is required for an anorexic ferret. Supplemental food should be made just wet enough to fit through the tip of a 6-ml syringe without clogging up or pouring through the tip. Recovering ferrets should be fed as much as they will consume every 3–4 hours. The slurry can be made with either a commercial carnivore diet or a canned prescription diet already available in the clinic or hospital. The food should be slightly warmed in a bowl of hot water. Warmed food has a more appealing smell and stimulates the patient's appetite. The food should never be heated in a microwave.

Blood Collection

UNIT SEVEN

OBJECTIVES

Upon completion of this unit, the reader should be able to:

▶ Perform venipuncture

© Eric Isselee/Shutterstock.com

GUIDELINES FOR BLOOD COLLECTION

Although blood may be collected from various sites, the lateral saphenous and cephalic veins provide ease of access for small amounts. Larger volumes may be collected from the right jugular and the anterior vena cava. If attempting an anterior vena cava blood collection, general anesthesia is recommended.

PURPOSE

- To collect an adequate volume for in-house blood and biochemistry analysis
- To collect and prepare a sample for an outside specialty exotic animal lab

COMPLICATIONS

- Inadequate restraint
- May require general anesthesia

EQUIPMENT

- 25–27 gauge needle
- 3 ml syringe
- Alcohol wipes
- Blood collection tubes
- Small-mammal face mask
- Anesthesia machine
- Confirmation from outside laboratory their requirements for any tests to be run
- Assistant for restraint

PROCEDURE for Collecting a Blood Sample from the Vena Cava

TECHNICAL ACTION	RATIONALE/AMPLIFICATION
1. Have all equipment ready.	
2. Confirm patient, tests, and sample requirements.	
3. Review guidelines for anesthesia of the ferret.	
4. Hold the patient with one hand, and gently place the mask entirely over the head.	
5. Introduce oxygen flow, then gas anesthesia.	
6. When a suitable plane of anesthesia is reached, lay the ferret on a padded table in dorsal recumbency.	
7. Have an assistant scruff the ferret so the head is back and the neck is slightly extended.	7a. The vena cava is located more caudally in ferrets than in other mammals.
8. The restrainer should extend both forelegs caudally.	
9. Wipe the puncture site with alcohol.	

(Continues)

TECHNICAL ACTION	RATIONALE/AMPLIFICATION
10. Insert the needle with syringe attached directly into the vena cava (**Figure 3-9**). When blood flashes in the hub, withdraw draw sample.	10a. 11a. These actions can be completed simultaneously and without delay.
11. Remove needle, apply pressure to puncture site until bleeding stops.	

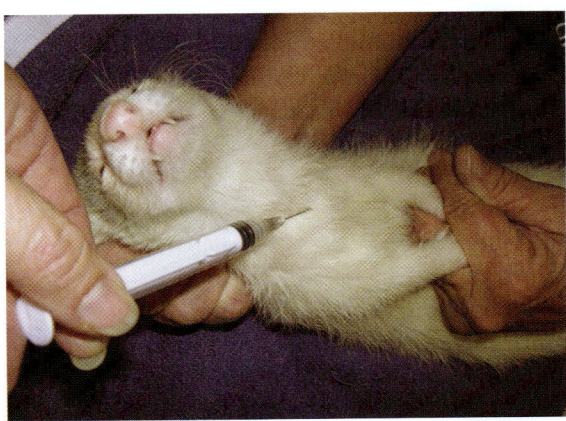

▲ **FIGURE 3-9** Positioning a ferret for a blood draw from the vena cava.

TECHNICAL ACTION	RATIONALE/AMPLIFICATION
12. Turn off gas anesthesia, allow a few breaths of pure oxygen.	
13. Recover the patient from anesthesia.	13a. Anesthesia time is usually under 5 minutes, and recovery from this quick mask-down should be rapid. The ferret should be held gently or placed in a readily observable enclosure.

UNIT EIGHT

Medication Administration

GUIDELINES FOR ADMINISTERING MEDICATIONS

Vaccinations should be administered subcutaneously (SQ). When held by the scruff, most ferrets will not object too much. The injection should be given under the skin just behind the scruff. Intramuscular (IM) injections are extremely painful and will elicit a loud scream from the ferret. If there is no alternative route, the drug can be injected in the quadriceps of the hind leg. It is imperative that a firm scruff be maintained and that the rear legs are also restrained. The ferret should be held low over a padded table so if it does break the scruff, it will not fall to the floor or hit the table hard.

Oral medications are accepted readily by the ferret if the tablets are ground and mixed with a small amount of soft carnivore diet blended to the same consistency as supplemental feeding and presented with a syringe. Alternatively, medications can be crushed and mixed into a small amount of meat-based human infant food. The one exception to this is metronizadole (Flagyl®); the taste is so bitter that no matter how it is disguised, the ferret will try to reject it with all it can muster—gagging, foaming at the mouth, clawing at the mouth, squirming, and contorting—trying to spit it all out. Patience and a good scruff are required.

✳ ADMINISTRATION OF RABIES VACCINATIONS

In most states, only a veterinarian can deliver a rabies vaccine. In a few states, a veterinary technician may deliver a rabies vaccine but only under the direct supervision of a veterinarian. Veterinary staff can draw up the rabies vaccination and have it ready for the veterinarian to administer.

PURPOSE

- To deliver vaccination against distemper.
- To deliver the prescribed therapeutic drug

COMPLICATIONS

- Pain associated with IM injections
- Medication refusal (metronizadole)

EQUIPMENT

- 25–27 gauge needle
- 1 ml or 3 ml syringe depending on volume
- Vaccine approved for use in ferrets
- Pill cutter or mortar and pestle for grinding tablets
- Soft carnivore diet or other flavored liquid to mix with medication
- Bowl of heated water for warming oral mixture
- Correct prescribed medication and dose
- Assistant for restraint if IM injection

PROCEDURE for Administering a SQ Injection in the Ferret

TECHNICAL ACTION	RATIONALE/AMPLIFICATION
1. Draw the prescribed dose into a 3-ml syringe with a 25–27 gauge needle.	
2. Hold the ferret by the scruff in one hand and administer the injection under the tent of skin just caudal to the scruff.	
3. If administering a distemper vaccination, the patient should remain in the clinic under staff observation for 20–30 minutes.	3a. Some ferrets have a vaccination reaction, which can be mild, severe, or life-threatening. If **anaphylaxis** occurs post-vaccination, it is usually within the timeframe above. Depending upon the severity of anaphylactic shock, antihistamines and/or short-acting steroids are administered. In severe cases, the ferret may have to be intubated and given respiratory support.

DISTEMPER VACCINATION

Even though the same distemper virus affects dogs, canine-specific "combo vaccines" should never be used in a ferret. The other components of the combo vaccine will cause severe adverse reactions in the ferret. Administer only a canine distemper vaccine approved for use in ferrets.

GUIDELINES FOR FLUID THERAPY

In critically ill ferrets, fluids can be delivered through an intraosseous (IO) catheter placed by the veterinarian. Preparation for placement of the catheter requires anesthesia and aseptic technique. Preferred sites for catheter placement are the femur, proximal end of the humerus, and wing of the ilium. Two complications may arise in using IO catheters: the rate at which fluids can be delivered, and the potential for introducing bacteria into the **intramedullary** space (the bone cavity) causing **osteomyelitis**, inflammation of the bone. The area should be clipped and a full surgical scrub performed. Aseptic technique must be maintained throughout the procedure. The person inducing and monitoring the anesthesia and recovery must be especially vigilant, as the ferret is likely to already be critical.

More frequently, veterinary technicians administer fluid therapy SQ. Typically, a ferret having a dehydration status of mild to moderate would receive 20–60 ml/kg/hour.

PURPOSE

- To restore and maintain hydration status

COMPLICATIONS

- None

EQUIPMENT

- 22-gauge needle
- Syringe adequate to contain volume of fluids to be delivered (6–12 ml)
- Extension set
- Prescribed fluids and amount to be administered

PROCEDURE for Administering SQ Fluids in the Ferret

TECHNICAL ACTION	RATIONALE/AMPLIFICATION
1. Draw up the correct amount of fluids into the syringe.	
2. Attach the extension set to the syringe tip.	
3. Attach the needle to the other end of the extension set.	
4. The restrainer scruffs and stretches the patient in lateral recumbency.	4a. This method allows the ferret some movement without pulling the needle out of the skin or driving it deeper into muscle tissue.
	4b. Refer to **Figure 3-10**.

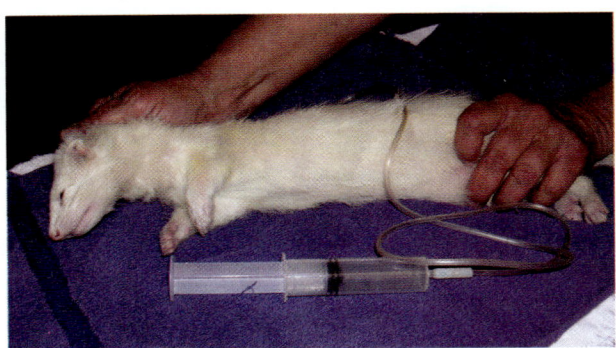

▲ **FIGURE 3-10** Administration of SQ fluids is made easier and is more comfortable for the ferret with the use of an extension set.

5. Administer fluids SQ, just caudal to the scapula.

UNIT NINE

Anesthesia

Upon completion of this unit, the reader should be able to

▶ Induce, monitor, and maintain anesthesia

GUIDELINES FOR ANESTHESIA OF THE FERRET

Ferrets respond well to inhalation anesthesia, as induction is smooth and recovery is usually rapid and uneventful. Most patients are induced by using a small-mammal face mask, which should be placed over the ferret's entire head. During induction, the mask can be held in place with one hand while the patient is held with the other hand. Ferrets rarely struggle during induction, so it is not necessary to maintain a scruff. The isoflurane induction rate is 2.5%. The oxygen flow rate is approximately 25–30 mL/kg/min. Anesthetic depth should be monitored during all procedures and the flow rate adjusted accordingly.

Intubating the ferret is easier than with many other species, as the epiglottis is easily visualized. An uncuffed 2.5 f ET tube is recommended. When a suitable plane of anesthesia is achieved, the ferret can be intubated. A length of gauze placed just behind the canine teeth aids in opening the mouth (**Figure 3-11**). This helps avoid a "clamp bite" if the ferret responds to the stimulus of tube placement. If the gauze is wet (damp), the threads are less likely to snag on the teeth. Once placed correctly, the ET tube is tied-in with another length of gauze in the same manner as with dogs and cats.

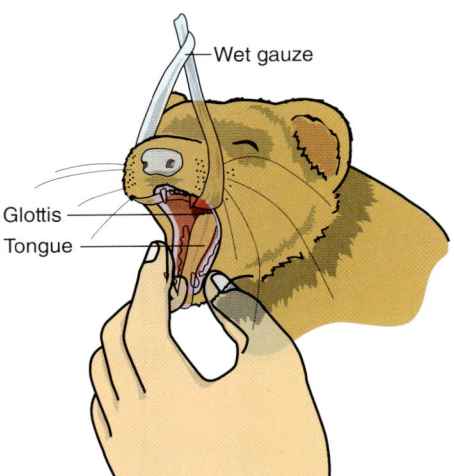

Wet gauze

Glottis

Tongue

▲ **FIGURE 3-11** A length of dampened gauze will assist in the intubation of a ferret. The gauze is used to hold the mouth open to visualize the trachea.

PURPOSE

- To deliver and maintain anesthesia for surgery
- To quickly mask down the patient for short procedures (jugular or vena cava blood collection
- To avoid causing pain or further trauma in positioning an injured patient or suspected abdominal blockage for radiographs

COMPLICATIONS

- Anesthesia depth is too deep or too light
- Surgical drape and size of patient make it more difficult to assess heart and respiration rate
- Reservoir bag attached to anesthesia machine is too large to adequately observe respiration rate

EQUIPMENT

- Standard small-animal anesthesia machine
- Small-mammal face mask
- Sterile eye lubricant
- Uncuffed endotracheal tube (2.5 f)
- Stethoscope
- Pulse oximeter

ESOPHAGEAL STETHOSCOPE

Monitoring ferrets' heart rate under anesthesia is sometimes difficult because of their small size and the surgical drapes used to keep the incision site sterile. One way to simplify monitoring the heart rate (HR) without interfering with the surgical field and risk of contamination is to use an esophageal stethoscope. Esophageal stethoscopes are quickly made by using a 22-inch red rubber catheter and a standard stethoscope. To do this, remove (pull off) the bell end of the stethoscope and replace it with the wide end of the red rubber catheter, in effect extending the reach of a normal stethoscope. Ensure the connection is secure.

When the patient is fully anesthetized, intubated, and tied into position on the surgical table, wet the catheter with warm water, shake off excess, and gently slide the catheter into the esophagus, over the tongue and just behind the ET tube. (Wetting the catheter first reduces friction as it introduced into the esophagus.) Slowly adjust the position of the catheter until it is positioned directly over the heart and clear heart sounds are audible. The patient then can be draped with no risk of contamination of the surgical site by having to disrupt the drapes repeatedly to obtain the HR.

PROCEDURE for Ferret Anesthesia

TECHNICAL ACTION	RATIONALE/AMPLIFICATION
1. Confirm the patient and procedure.	
2. Place a small amount of sterile lubricant in each eye.	2a. This prevents the eyes from drying out. The ointment tube tip should not touch the eye. Use only a small amount, one-fifth of an inch strip, of ointment, and place it medial to lateral onto the cornea. The normal blink reflex will cause the ointment to move across the eye.
3. Hold the ferret in one hand.	
4. Place a small-mammal face mask entirely over the patient's head.	
5. Turn on the oxygen, and allow the ferret to have four or five breaths of pure oxygen.	

(Continues)

TECHNICAL ACTION	RATIONALE/AMPLIFICATION
6. Turn the vaporizer to 2.5%.	
7. Hold the patient and face mask until a suitable plane of anesthesia is achieved.	
8. Have an assistant hold open the ferret's mouth.	8a. Refer to Figure 3-11.
9. Visualize the glottis, wait for a breath, and intubate.	
10. Transfer the patient to the anesthesia machine.	
11. Record heart and respiration rates (place the esophageal stethoscope when the patient is in position in the OR).	11a. When the patient is tied into position on the OR table, the esophageal stethoscope is easily placed and maintained in position over the heart and adjusted if necessary.
12. Support with positive pressure ventilation (PPV) if necessary.	
13. Prepare the patient for surgical procedure.	

Surgical Preparation

OBJECTIVES

Upon completion of this unit, the reader should be able to

▶ Prepare patient for surgery

GUIDELINES FOR SURGICAL PREPARATION

Preparation of the patient for a surgical procedure is no different than that for a dog or cat. The surgical field should be shaved using a #40 blade, and scrubbed using aseptic technique. The procedure will determine the area to be shaved, and the same field parameters used in other small-animal surgeries are used for the ferret.

Ferrets should not be fasted for more than one or two hours because of the risk of hypoglycemia if the patient is already compromised.

Prior to inducing the ferret, the surgical table should be set up with a recirculating water pad so it is warmed before placing the patient on the OR table.

Change the tie-in ropes on the OR table to lengths of gauze. The ropes (or cords) used for dogs and cats usually are too bulky and harsh for the small extremities of the ferret.

PURPOSE

- To prepare the patient for surgery

COMPLICATIONS

- Dull blade on clippers
- Lack of preparation

EQUIPMENT

- Anesthesia machine
- Small-mammal face mask
- Length of gauze
- Uncuffed ET tube
- Disposable gloves
- Electric clippers and #40 blade
- Scrub basin and gauze pads
- Stethoscope
- Red rubber catheter for use as esophageal stethoscope
- Anesthesia record sheet, clipboard, and pen
- Surgical attire: cap and mask

PROCEDURE for Surgical Preparation of the Ferret

TECHNICAL ACTION	RATIONALE/AMPLIFICATION
1. Induce anesthesia using the face mask.	
2. Intubate and transfer the patient to the anesthesia machine.	
3. Shave the surgical field, don gloves, and perform a surgical scrub of the area.	

(Continues)

TECHNICAL ACTION	RATIONALE/AMPLIFICATION
4. Turn off the anesthesia machine in the prep area, and transfer the patient to the OR.	4a. 5a. Smaller facilities may have only one portable anesthesia machine.
5. Reconnect the patient to the anesthesia machine in the OR (or have assistant re-locate the machine from the prep area to the OR).	
6. Check that the heating pad is functioning and warm.	
7. Position the patient on the surgical table, and tie-in with lengths of gauze.	7a. Typically, tie-downs attached to the OR table are cord or rope; these are too heavy to use with a small patient.
8. Monitor anesthesia: depth, heart, and respiration.	
9. If using an esophageal stethoscope, change the stethoscope bell and replace it with a rubber catheter.	9a. The bell of the stethoscope pulls off easily; replace it with the larger end of the red rubber catheter (**Figure 3-12**).

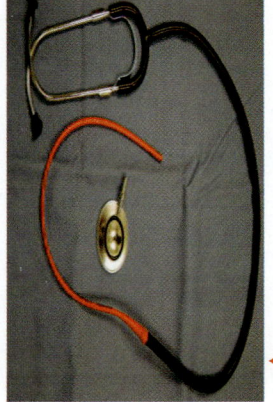

◀ **FIGURE 3-12** Use of the esophageal stethoscope facilitates obtaining an HR in the ferret during surgery.

TECHNICAL ACTION	RATIONALE/AMPLIFICATION
10. Place the esophageal stethoscope and confirm audible heart sounds.	10a. Wet the catheter before placing it into position in the esophagus so it will slide into the esophagus without causing irritation. For measurement, locate the heart with your fingers and measure approximate distance so the tip of the catheter, when inserted, is positioned over the top of the heart.
	10b. Slide the catheter behind the endotracheal tube and into the esophagus until clear heart sounds are heard. The heart is located more caudally in the ferret than in other species, but take great care not to enter the stomach. Auscultate with the ear pieces to find the correct location. Once in place, only minor adjustment might be needed during the surgery.
11. Connect other monitoring devices that will be used.	
12. Put on your own cap and mask.	
13. Do not leave patient; continue to monitor anesthesia using the Anesthesia Record sheet.	

Radiology

OBJECTIVES

Upon completion of this unit, the reader should be able to:

▶ Take radiographs, ventral/dorsal (V/D) and lateral

© Eric Isselee/Shutterstock.com

GUIDELINES FOR TAKING RADIOGRAPHS OF THE FERRET

The two views normally taken in the ferret are the same as for other species: lateral and V/D (**Figure 3-13A and B**). A ferret should be positioned the same as a dog or a cat. Settings on the X-ray machine are comparable to a small cat or a rabbit.

A well-maintained scruff is usually all that is required to restrain the ferret for radiographs. Follow radiation safety measures, and make sure that the restrainer's hands are shielded.

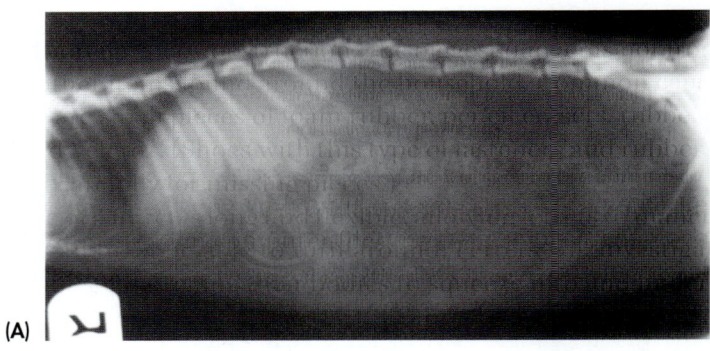

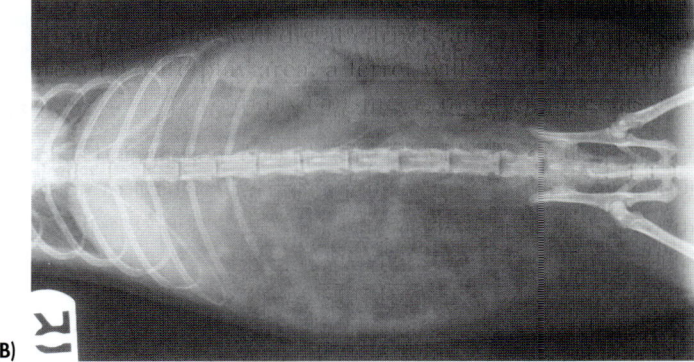

▲ **FIGURE 3-13** (A) Lateral view of the ferret abdomen (B) V/D view of the ferret abdomen.

PROCEDURE for Taking Radiographs of a Ferret

TECHNICAL ACTION

1. Have cassettes or digital plates ready, one for each view to be taken.

2. Imprint card should be filled out with the correct patient details.

3. Wear radiation safety apron, thyroid shield, and glasses.

4. Scruff and position the patient.

5. Have the assistant cover the restrainer's bare hands.

(Continues)

TECHNICAL ACTION

6. Back away from the beam as far as possible while maintaining the scruff.

7. Use the pedal remote to fire the X-ray beam.

8. Reposition the ferret and repeat steps 4–7.

FAST FACTS

Weight
- Hobs: 1–2 kg (2.2–4.2 pounds)
- Jills: 0.6–1 kg (1.3–2.2 pounds)

Vital Statistics
- Temperature: 37.7°–39.7° C (100°–103.5° F)
- Heart Rate: 180–250/minute
- Respiration rate: 33–36/minute

Vaccinations
- Canine distemper: Vaccination schedule: 6–8 weeks; 9–12 weeks; 14–16 weeks; yearly thereafter
- Rabies: Vaccination schedule: 4–6 months; yearly thereafter

Life Span
- 8–11 years

Reproduction
- **Sexual Maturity:** 6–12 months
- **Gestation:** 41–43 days
- **Litter size:** 1–18 (8 average)
- **Weaning age:** 6–8 weeks

Zoonotic Potential
- **Bacterial:** Salmonella
- **Protozan:** Giardia, Cryptosporidium, Toxoplasma
- **Fungal:** Ringworm

Viral
- Rabies: All mammals can potentially carry and transmit rabies. Because of the vaccination requirement and limited potential of exposure, it is extremely rare in domestic ferrets.

Review Questions

1. What are three other species closely related to the ferret?

2. Ferrets should be given food "free choice." What does this mean, and why is it so important to the ferret?

3. What is the result if a ferret is bathed too frequently?

4. Why is it important to ask the client if anyone in the household has had the flu?

5. What is the vaccination protocol for ferrets?

6. What are the laws in your state, county, and municipality regarding the possession of ferrets?

7. What is the consequence if a jill is allowed to remain in estrus without being bred?

8. Describe the normal ear exudate and how it can be determined if a ferret has ear mites.

9. Why is alopecia a significant finding on the physical examination?

10. An anorexic ferret should be fed every —— hours. How much should be given at each feeding?

11. Why should a ferret remain in the clinic for 20–30 minutes after a vaccination?

12. What are two complications of delivering fluids IO?

13. Describe an esophageal stethoscope and what is the benefit of using this during surgery?

14. What is a mast cell tumor?

15. What is the significance of two tattooed dots in a ferret's ear?

References

Judah, Vicki, & Nuttall, Kathy *Exotic Animal Care & Management* (2005). Clifton Park, NY: Cengage Learning.

Quesenberry, K.E., & Orcutt, C. *Basic Approach to Veterinary Care* (2004). In K.E. Quesenberry & J.W. Carpenter (Eds.). *Ferrets, Rabbits and Rodents: Clinical Medicine and Surgery,* 2nd ed. (pp. 58–71) Imprint of Elsevier Science, St. Louis, MO.

Sager, William C. *American Ferret Report* Vol. 9, No. 25 (1998), pp. 19–21 (accessed June 23, 2012)

www.http://kingcounty.gov (accessed July 28, 2012)

CHAPTER 4

GUINEA PIGS

An animal's eyes have the power
to speak a great language.

— Martin Buber

UNIT
ONE

Overview of Species

OBJECTIVES

Upon completion of this chapter, the reader should be able to:

▶ Understand the basics of guinea pig behavior

KEY TERMS

ascorbic acid	precocial
boar	pubic symphysis
cavy	pup
drilling	scurvy
ossify	sow
polyestrous	

GUINEA PIGS

The guinea pig, technically known as the **cavy** (plural, cavies), is a popular pet for children and adults (**Figure 4-1**). Guinea pigs are known for their gentle nature and for being social with people and each other. Their eyesight is poor, but their hearing is highly developed. Familiar footfalls of the owner arriving home often elicit a variety of whistles, chirps, and soft chuckling sounds.

▲ **FIGURE 4-1** Guinea pigs are popular pets for children.

BEHAVIOR

Guinea pigs produce a wide range of vocal sounds. The whistle can be a call of greeting or of alarm, the exact meaning determined by the length and pitch of the call. Owners closely involved with their guinea pigs are not only able to interpret the situation but also know which cavy in the group is whistling. Chirps and chuckles are lower-pitched and equally varied in tone and modulation, and these are sounds of contentment. Distinct from other sounds is **drilling,** a low, almost guttural and rapid d-r-r-r-r sound produced by the teeth. Drilling is a warning, or an alert to other guinea pigs of possible danger. Also, a **boar** (male), in pursuit of a **sow** (female) will drill to challenge other males. Drilling is also heard when a guinea pig is in pain and never should be interpreted as a contented purr.

Guinea pigs have little in the way of self-defense. Their oral cavity is small, which makes it difficult for them to bite. They have claws on all four feet, and scratches may be likely if they do not feel secure when they are held. The normal group (or herd) response to fear is to flee, to stampede. Often, a single guinea pig will freeze in place, not moving until the perceived danger has passed. Attempts to capture a loose guinea pig can be highly stressful to the animal. It is best to let it settle, then tempt it to approach by offering food. In most instances, the sound of fresh crunchy lettuce or the scrunch of a cellophane bag will convince the guinea pig to approach. If handled regularly, it then can be scooped up in two hands, one around the body and the other supporting the rump (**Figure 4-2**).

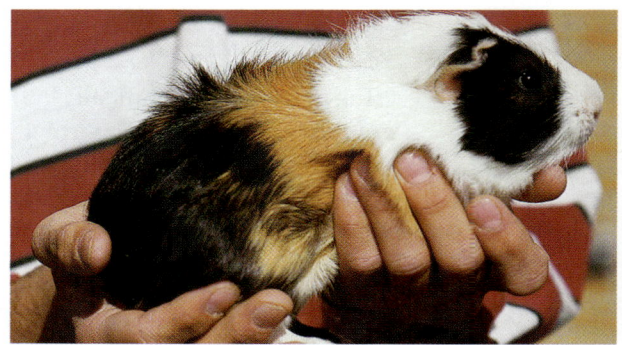

▲ **FIGURE 4-2** Guinea pigs feel more secure when they are held with both hands and the rump is supported.

Cavies can be housed simply. Cages with platforms and wire floors are not suitable. They do not climb or attempt to jump out, and any enclosed area with a smooth floor and sides of around 8–10 inches will safely contain them. The floor space should be at least 36 inches long and provide ample room for a hide box and food bowls, and have a method for attaching a water bottle (glass aquariums are not suitable because there is little air circulation and glass magnifies heat). An enclosure with a clamp-on wire top is ideal because it prevents other pets from having direct access to the guinea pig.

Guinea pigs urinate and defecate without preference and frequently contaminate food and water bowls. Using a food hopper and water sipper bottle prevent this from occurring. Guinea pigs drink a lot of water and need to be provided with a sipper bottle that meets their daily needs. Bottles should be cleaned out and refilled with fresh water daily. Guinea pigs often drink when they have mouthfuls of food, play with the sipper tube, and frequently push food material into the tube, where it swells and completely blocks the flow of water.

BREEDING

Determining the sex of a cavy is fairly easy compared to other small rodents. The anogenital area of sows is more rounded in appearance and has a Y shape. The Y configuration extends from the anus to the vaginal opening (**Figure 4-3**). Boars have prominent, slender testicles on either side of the anal area. However, because of their open inguinal canal, the testes are not always apparent, so young guinea

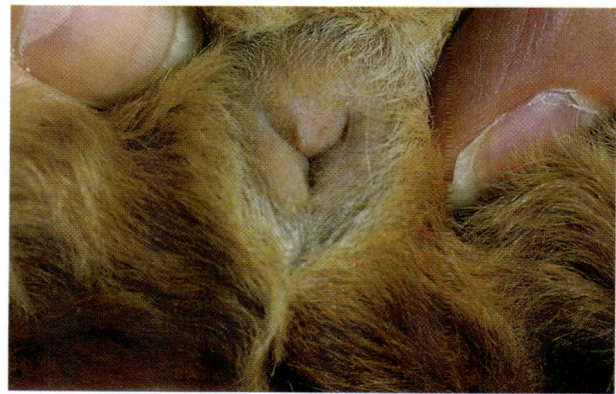

▲ **FIGURE 4-3** Genitalia of a female guinea pig.

pig **pups** may be sexed incorrectly. Applying gentle pressure with the thumb and a caudally directed stroke on the abdomen will cause the penis to protrude and confirm the sex (**Figure 4-4**).

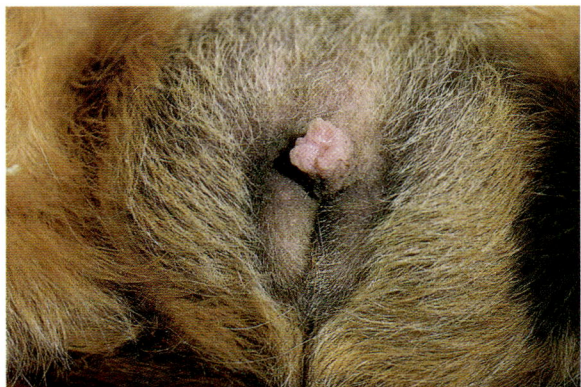

▲ **FIGURE 4-4** Genitalia of a male guinea pig.

Early determination of the sex is necessary to avoid unwanted litters. Boars reach sexual maturity around 9–10 weeks. Sows become sexually mature earlier, around 4–6 weeks.

Sows are **polyestrous** in that they have many breeding cycles throughout the year. If a sow does not have her first litter before she is 7–9 months old, the **pubic symphysis** (cartilaginous pelvic midline), which normally relaxes prior to parturition, begins to **ossify**, or fuse, preventing normal delivery. In those cases, Caesarian delivery is necessary to prevent the death of the sow and her litter.

Guinea pigs are born **precocial** after a gestation period of 63 days. From birth, their eyes and ears are open, they have a full coat of hair, and their teeth are erupted. They are highly active from birth and start to nibble at hay and the fecal pellets of adults when they are a few days old. Guinea pigs are coprophagic, and ingestion of adult fecal material is necessary for the pups to help establish normal gut flora.

Both sexes have a single pair of inguinal nipples. Despite having only two mammary glands, sows are able to take care of an average-sized litter of three or four pups. Unweaned pups patiently await their turn to nurse and show no aggression toward one another. It is not necessary to remove the boar from the presence of the neonates. Boars are tolerant of their antics and will often groom the pups.

DIET

Cavies are herbivores and can be fed a variety of fresh greens and vegetables. Daily offerings should include guinea pig pellets supplemented with timothy or other grass hay. Although appearing to be the same as rabbit pellets, guinea pig pellets have added **ascorbic acid**, vitamin C. Ascorbic acid is an essential part of guinea pigs' daily diet, as they are unable to synthesize it from other foodstuffs. Ascorbic acid is necessary for the metabolism of cholesterol, amino acids, and carbohydrates. Without adequate daily amounts in the diet, the signs of vitamin C deficiency can appear in as few as 10 days. If immediate veterinary care is not received and the diet corrected, the guinea pig will die from a disease called **scurvy**, which is easily prevented by feeding the correct diet.

Foods high in vitamin C and readily consumed include kale, parsley, beet greens, spinach, and broccoli. Once introduced, guinea pigs will also eat red and yellow peppers, tomatoes, kiwi fruit, and orange segments. Vitamin C supplements are available as drops or tablets, and some are added to the water. The latter can create problems because many guinea pigs refuse to drink the treated water. Also, ascorbic acid is a water-soluble vitamin and quickly degrades in water.

Generally, guinea pigs are hardy and easy to care for. However, the veterinary technician should become familiar with the unique medical concerns and specific dietary and husbandry needs of the guinea pig.

Restraint Techniques

OBJECTIVES

Upon completion of this unit, the reader should be able to:

▶ Correctly restrain the guinea pig

GUIDELINES FOR RESTRAINT OF THE GUINEA PIG

Guinea pigs are easy to handle and are often vocal when they are picked up. Unless fully supported, they may struggle in an attempt to "find their footing" rather than an attempt to escape. Holding them securely with both hands makes them feel more secure. They have little scruff, and attempting to use this method of restraint will only cause alarm and pain, as it results in more of a severe pinch than a scruff.

PURPOSE

- To hold for an examination
- To restrain for a medical procedure or treatment

COMPLICATIONS

- None

EQUIPMENT

- Towel or non-skid mat on a slippery surface

PROCEDURE for Restraint of the Guinea Pig

TECHNICAL ACTION	RATIONALE/AMPLIFICATION
1. Put a towel or non-skid mat on the examination table.	
2. Place the securely held patient on the table and cup the guinea pig between your hands (**Figure 4-5**).	2a. Refer to Figure 4-5.

▲ **FIGURE 4-5** To restrain a guinea pig for an examination, the patient should be placed on a towel and cupped in both hands.

Grooming

OBJECTIVES

Upon completion of this unit, the reader should be able to:

▶ Perform nail trims, and remove anal impactions

GUIDELINES FOR GROOMING THE GUINEA PIG

Guinea pigs have four toes on the front feet and three toes on the hind feet. Guinea pig nails grow continuously and should be checked often. Because guinea pigs usually are kept on a soft substrate, the nails are not worn down naturally. In severe cases, the toenails may curl into the soft foot pads or become very long and twisted, making it difficult for the guinea pig to walk normally.

It usually is not necessary to bathe a guinea pig unless it has become cage-soiled or stained from an episode of diarrhea. If a complete bath is necessary, a mild kitten shampoo should be used, with warm and shallow water. Water should be poured over the guinea pig in a small tub, shampooed, then rinsed completely, and towel-dried. The guinea pig should not become chilled and should be placed in a warm area until completely dry.

More frequently, the rump area is cleaned and bathed rather than giving guinea pigs a complete bath. They have large peri-anal sebaceous glands used for scent marking, and these glands are especially pronounced in unneutered boars. They drag their rumps, pressing down on the floor surface to leave a scent trail. When excessive marking occurs in males, a buildup of natural skin oils coats the hair around the glands, giving the hair a greasy appearance. The scent glands and folds of skin around the genitals can easily become impacted with fecal and bedding material, resulting in a foul smell. The affected area has to be cleaned and the impaction removed.

PURPOSE

- To trim nails to the correct length
- To clean soiled areas of the body
- To treat peri-anal gland impaction

COMPLICATIONS

- None

EQUIPMENT

- Small cat claw trimmers or human nail clippers
- Mild kitten shampoo
- Shallow tub for bathing
- Basin
- Mineral oil
- Cotton-tipped swabs

PROCEDURE for a Nail Trim

TECHNICAL ACTION	RATIONALE/AMPLIFICATION
1. Isolate each foot.	1a. Refer to Figure 4-6.
2. Use small cat claw trimmers or human nail clippers to trim each nail (**Figure 4-6**).	

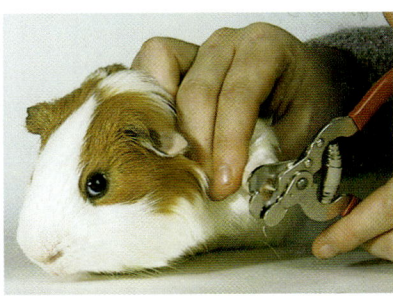

▲ **FIGURE 4-6** Guinea pigs' nails can be trimmed using a cat claw trimmer.

PROCEDURE for Cleaning Impacted Peri-anal Glands

TECHNICAL ACTION	RATIONALE/AMPLIFICATION
1. Apply mineral oil with cotton-tipped swabs to help soften the impacted material.	
2. When the impaction is removed, inspect the affected area closely.	2a. Look for areas of redness, ulceration, or any abnormality of the skin.
	2b. If sores or ulcerations are present, do *not* apply any topical medication without approval from the treating veterinarian. Guinea pigs do not tolerate many of the commonly used antibiotics. They may lick off and ingest topical creams. (Refer to Guidelines for the Administration of Medications.)
3. Place the guinea pig in a shallow tub and use the basin to wet the rump area.	
4. Use warm water and kitten shampoo to bathe and remove residual odor.	
5. Thoroughly dry the guinea pig with a towel.	

Patient History

OBJECTIVES

Upon completion of this unit, the reader should be able to:

▶ Complete an accurate patient history

KEY TERMS

substrate

GUIDELINES FOR TAKING A PATIENT HISTORY

The standard small-animal history form is used. Specific details should be noted regarding diet, husbandry, breeding status, and other pets in the household. Although they are generally hardy little animals, most problems can be directly related to issues of diet, habitat, reproductive concerns in the sow, and stress.

PURPOSE

- To obtain as much information as possible to assist in a diagnosis

COMPLICATIONS

- Complete history may not be known

EQUIPMENT

- Small-animal examination form
- Patient computerized file or paper copy of exam form
- Pen

PROCEDURE for Taking a Patient History

TECHNICAL ACTION	RATIONALE/AMPLIFICATION
1. Begin with the chief complaint.	
2. Complete the signalment, and progress through each section of the form.	
3. Gather as much information as possible regarding diet and any supplements given.	3a. Guinea pigs need a daily intake of vitamin C.
4. Ask the client to describe the habitat, including the floor area, number of guinea pigs housed, **substrate** (type of bedding), type of container or enclosure, water source, and food bowls.	4a. Owners should be alerted to monitor water intake carefully, as guinea pigs can become dehydrated quickly.
	4b. Cedar bedding should never be used to house small mammals. When cedar shavings become wet, they release toxic fumes that cause severe respiratory problems and can lead to death. Aspen shavings or plain newspaper are both suitable bedding material.
5. Determine how and when the guinea pig was acquired.	5a. Although they are socially compatible as species, rabbits and guinea pigs should never be housed together. Rabbits can be asymptomatic carriers of *Bordatella bronchiseptica*, the same bacteria that is responsible for kennel cough in dogs. In guinea pigs, it causes severe respiratory tract infections and is responsible for abortions and still-births in pregnant sows. The mortality rate is high.
6. Inquire about other household pets.	

UNIT FIVE

Physical Examination

OBJECTIVES

Upon completion of this unit, the reader should be able to:

▶ Perform a physical examination

KEY TERMS

anorexia

GUIDELINES FOR PHYSICAL EXAMINATION OF THE GUINEA PIG

The standard small-animal physical examination form is used, and no areas should be left blank. By becoming knowledgeable regarding some of the specific concerns with guinea pigs, the findings, in combination with the patient history, can better assist in a diagnosis. Stress is a major contributor to health problems commonly seen in guinea pigs. For example, stress can be one cause of **anorexia**. With the guinea pig, anorexia is more than simply going off feed; it leads to major medical concerns and mortality.

PURPOSE

- To determine the general health status and record objective findings

COMPLICATIONS

- Small oral cavity, which may require specialized oral/dental equipment

EQUIPMENT

- Pediatric stethoscope
- Gram scale with tray
- Lubricating gel
- Thermometer
- Partial dental pack containing cheek spreaders and mouth gag*

PROCEDURE for Physical Examination of the Guinea Pig

TECHNICAL ACTION	RATIONALE/AMPLIFICATION
1. Proceed through the physical examination form.	
2. Carefully examine the head, submandibular area, and neck.	2a. Guinea pigs often present with swollen areas, lumps, in the submandibular region and neck. They may develop abscesses of the lymph glands or a localized pocket of infection because of oral trauma. Guinea pigs chew on bedding and rough stalks of hay, which may penetrate the oral mucosa and form an abscess.
3. Check the alignment of the incisors and note any signs of excess salivation.	3a. Incisor alignment is easily checked, but more frequently the lower premolars overgrow, irritating the gingiva and causing excess salivation. Because of the small oral cavity, this condition can be confirmed only with the use of a mouth speculum and cheek spreaders designed specifically for use in rodents and rabbits (**Figure 4-7***). If a complete examination of the oral cavity is required, most often the patient is placed under a light anesthesia (masked down) so as not to stress the patient. If overgrown premolars or molars are evident, they are clipped back while the patient is under a light inhalant anesthesia. This procedure normally is not performed as part of the physical exam; the patient usually is admitted.

(Continues)

TECHNICAL ACTION   **RATIONALE/AMPLIFICATION**

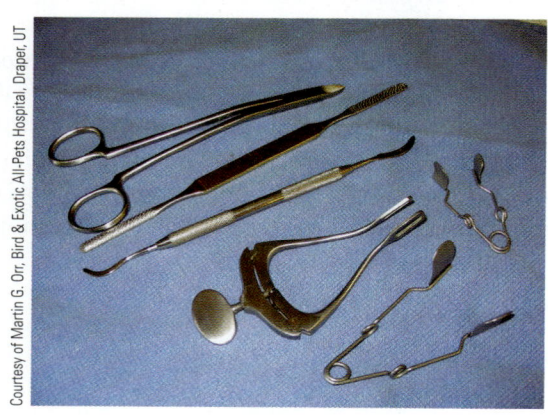

Courtesy of Martin G. Orr, Bird & Exotic All-Pets Hospital, Draper, UT

◀ **FIGURE 4-7** A complete oral examination requires the use of specialized equipment. Shown here are dental nippers, a tooth rasp, and a dental probe. There are two sizes of cheek spreaders and a mouth gag. One end of the mouth gag is placed over the top incisors, and the other is placed over the bottom incisors. The thumb screw is used to carefully open the mouth and hold the oral cavity open for a more detailed examination.

4. Observe the body shape for symmetry.

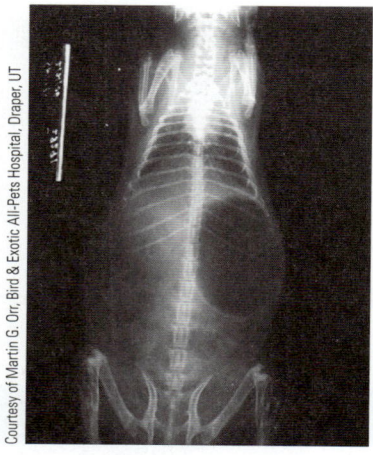

Courtesy of Martin G. Orr, Bird & Exotic All-Pets Hospital, Draper, UT

▲ **FIGURE 4-8** A dorsal/ventral (D/V) radiograph showing a large air pocket. The patient is sitting directly on film cassette without the use of anesthesia. This is possible because of the "freeze" behavior when guinea pigs are in strange surroundings.

5. Carefully palpate any abnormality or lumps found.

6. Auscultate the heart and lungs.

7. Examine each foot and nail growth.

8. Look for signs of impaction and debris in the peri-anal region.

9. Obtain an accurate weight in grams.

10. Lubricate the tip of a thermometer and gently insert into the rectum to obtain a temperature.

4a. Guinea pigs have a well developed and long gastrointestinal tract, and sometimes large air pockets form in the gut. If they are big enough, air pockets interfere with normal intestinal placement, and body asymmetry can be observed with abdominal distention (**Figure 4-8**). Large air pockets also cause difficulty breathing and can affect cardiac function.

7a. Guinea pigs kept on wire can develop pododermatitis and ulcers on the plantar surface of the feet.

Care of the Hospitalized Patient

OBJECTIVES

Upon completion of this unit, the reader should be able to:

▸ Care for the hospitalized patient

GUIDELINES FOR CARE OF THE HOSPITALIZED PATIENT

Guinea pigs should be housed away from predatory species in a quiet area of the hospital where they can be easily observed. It is important not to house them in the proximity of rabbits, as rabbits may be asymptomatic carriers of *Bordatella bronchiseptica*. In guinea pigs, *Bordatella* can become a life-threatening respiratory infection, especially in an already compromised cavy.

Guinea pigs may be confined in a small stainless steel kennel or a solid small-animal carrier, and the floor can be covered in plain newspaper. A sipper bottle should be provided and a heavy crock for pellets. Grass hay and fresh greens should always be available. A small cardboard box with one end cut away provides an inexpensive and disposable hide box.

Blood Collection

OBJECTIVES

Upon completion of this unit, the reader should be able to:

▶ Perform a blood collection

KEY TERMS

Kurloff bodies

GUIDELINES FOR BLOOD COLLECTION

No more than 0.5–0.7 ml/g of body weight should be collected. Prior to collecting a blood sample, always obtain an accurate weight in grams. One collection site frequently used is the lateral saphenous, as it causes the least amount of stress for the patient. For a greater blood volume the jugular vein can be accessed. Because of the restraint complications and stress to the patient, the cavy should be briefly masked down with inhalent anesthesia.

PURPOSE

- To obtain an adequate volume of blood for diagnostic panels and blood film examination

COMPLICATIONS

- Small, fragile veins that collapse easily
- Patient stress

EQUIPMENT

- 25-gauge needle
- Tuberculin or insulin syringe
- Microtainer collection tubes

PROCEDURE for Blood Collection

TECHNICAL ACTION	RATIONALE/AMPLIFICATION
1. The restrainer should cup the patient with all four feet on the table.	
2. The person performing the blood draw decides which vein to approach and restrains the limb.	
3. Isolate the vein with an alcohol-soaked cotton ball, and occlude the vessel above the puncture site.	
4. Enter the vessel and withdraw a sample.	4a. This is easier when the angle of the approach to the vein is decreased or by slightly bending the needle.
5. Release pressure on the vein and withdraw the needle.	5a. The sample should be immediately transferred to microtainer tubes and rocked gently.
6. Apply pressure to the puncture site.	

KURLOFF BODIES

Examination of the blood film should include the recognition and recording of **Kurloff bodies**, which are cytoplasmic inclusions in some monocytes. They are normal in the guinea pig and may increase in number and become more obvious in a pregnant sow. The function of Kurloff bodies is unknown.

Medication Administration

OBJECTIVES

Upon completion of this unit, the reader should be able to:

▶ Administer medications and understand antibiotic protocols specific to the guinea pig

KEY TERMS

endotoxemia

© Eric Lsselee/Shutterstock.com

227

GUIDELINES FOR THE ADMINISTRATION OF MEDICATIONS

Antibiotic choices in the guinea pig are limited. Therapy protocol, as always, should be directed by the veterinarian. Guinea pigs on antibiotic therapy should have a bio-chem panel during and after the treatment to monitor potentially lethal changes that may occur.

PURPOSE

- To achieve a therapeutic dose of prescribed medication

COMPLICATIONS

- Adverse reaction
- Disruption of normal gut flora
- Gut stasis
- Overgrowth of Gram-negative bacteria
- **Endotoxemia**
- Death

EQUIPMENT

- 25-gauge needle
- 1–3 ml syringe
- Alcohol swabs

PROCEDURE for an Intramuscular Injection

TECHNICAL ACTION	RATIONALE/AMPLIFICATION
1. Withdraw the prescribed amount of injectable medication.	
2. Restrain the patient on a solid surface.	2a. Tuck the patient securely, head first, into the crook of your arm. The patient is likely to whistle loudly and attempt to flee when the needle is introduced.
3. Swab the injection site with an alcohol-soaked cotton ball, separating coat.	
4. Inject medication into the quadriceps of the hind leg.	
5. Gently massage the injection site to help disperse the bolus and alleviate pain.	

PROCEDURE for Administering Oral Medication

TECHNICAL ACTION	RATIONALE/AMPLIFICATION
1. Prepare the prescribed dose.	1a. A flavoring may be added to make the medicine more palatable.
2. Restrain the patient with the rear quarters tucked into the crook of your arm.	
3. Gently open the mouth with the tip of the syringe.	
4. Administer the dose from the side of the mouth, directed toward the tongue. Deliver in small increments, and allow the patient to swallow (**Figure 4-9**).	4a. The guinea pig will not swallow the medication if it is deposited in the cheek. The approach is the same as used with rabbits.

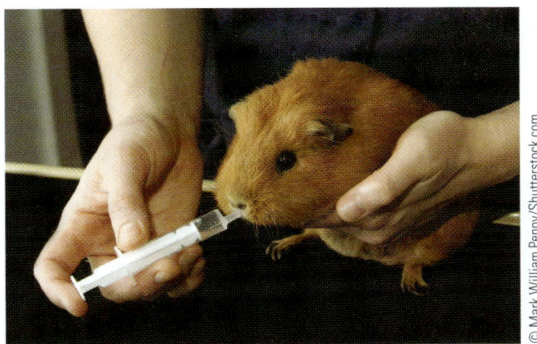

▲ **FIGURE 4-9** Administering oral medication to the guinea pig with the use of s syringe.

GUIDELINES FOR FLUID THERAPY

Peripheral catheters are difficult to place in guinea pigs because of their small, fragile vessels. Subcutaneous fluids can be given between the scapulas. The skin is tough, tight, and with little subcutaneous space. An intrascapular fat pad also makes the procedure more painful and slows the absorption rate. The patient is likely to vocalize loudly. Intraosseous catheter placement is often attempted; however, because of the inherent risks of anesthesia necessary for this procedure, it must be considered carefully. The anesthesia risk and difficulties of catheter maintenance usually make this an impractical option. An alternative site for a small bolus of subcutaneous fluids is the flank area, lateral to the spine.

PURPOSE

- To restore and maintain hydration status

COMPLICATIONS

- Tight skin with little subcutaneous space
- Slow absorption rate because of fat pad
- Pain and stress to the patient

EQUIPMENT

- Prescribed volume and type of fluids
- 25–27 gauge needle
- 3 ml syringe

PROCEDURE for Administering Subcutaneous Fluids

TECHNICAL ACTION	RATIONALE/AMPLIFICATION
1. Draw the prescribed amount of fluids into the syringe.	1a. Prior to administration, the fluids should be warmed to body temperature (37.2°–39.6° C; 99°–103° F)
	1b. Many clinics and hospitals that regularly treat small mammals maintain fluids in an incubator, or the fluids can be warmed by placing the syringe of fluids in a basin of heated water.
2. Restrain patient with all four feet on a solid surface.	
3. Administer fluids in small volumes.	
4. Record amount actually delivered with each injection and note the injection site.	4a. Because of the small amount of SQ space available, it is not always possible or prudent to attempt delivery of the entire amount of prescribed fluids at one time.
	4b. A large bolus of fluids in one site may cause such SQ pressure that the fluids will dribble out from the puncture site.
	4c. Using alternate sites reduces the pain and causes less trauma and disruption to underlying tissue integrity.

Anesthesia

OBJECTIVES

Upon completion of this unit, the reader should be able to:

▶ Explain anesthesia procedures and complications specific to the guinea pig

GUIDELINES FOR ANESTHESIA OF THE GUINEA PIG

Guinea pigs are extremely difficult to intubate because of their small oral cavity and difficulty in visualizing the trachea. They should be masked down with either isoflurane or sevoflurane. Guinea pigs should not be fasted for more than 2 hours prior to anesthesia. Though they lack the ability to vomit, regurgitation during anesthesia is common and may be frequent. As soon as a light plane of anesthesia is achieved, the oral cavity must be swabbed out, removing food and debris from their cheeks and molars. This may be necessary several times during the period of anesthesia to prevent aspiration.

Recovery should take place in a quiet area with unobtrusive observation. Guinea pigs sometimes fail to recover because of dramatic metabolic changes that can occur during the anesthesia period and the stress of induction. Rapid induction reduces stress. Guinea pigs with any medical concerns should be handled minimally because of the potentially catastrophic results of stress.

Anesthesia usually is administered only when there is no other option, for example, when a cesarean delivery is required to save the lives of the litter. In this unfortunate situation, the sow's survival rate is low and the usual objective is to deliver a viable litter.

CARE OF THE ORPHANED LITTER

Fecal pellets from a healthy guinea pig are sometimes mixed with a supportive diet and fed by syringe to an orphaned litter. "Coprophagic therapy" has been used incidentally and with some success; however, the risk of introducing salmonella to a neonate has to be carefully considered and weighed against the normal ingestion of fecal pellets. Alternatively, another sow with a nursing litter may adopt the orphans. Care must be taken to observe the sow and monitor her health and nutrition so she does not become overwhelmed with the demands of a second litter. Fortunately, guinea pigs will begin to eat solid food within a day or two after giving birth.

PURPOSE

- To deliver anesthesia for a cesarean delivery
- To castrate a boar
- To use the only alternative to provide treatment of severe injuries, abscesses, or fractures
- To apply pre-euthanasia protocol

COMPLICATIONS

- Stress to the patient
- Dramatic metabolic changes
- Uncertain recovery
- Death

EQUIPMENT

- Standard small-animal anesthesia machine with non-re-breathing system
- Small-animal face mask
- Eye lubricant
- Quantity of cotton-tipped swabs
- Pediatric stethoscope

PROCEDURE for Anesthesia in the Guinea Pig

TECHNICAL ACTION	RATIONALE/AMPLIFICATION
1. Restrain the patient.	
2. Apply a small amount of lubricant to each eye.	
3. Flush the anesthesia system with 100% oxygen.	
4. Gently introduce the face mask, and position it over the entire head.	
5. Introduce anesthesia gas: isoflurane at 5.0%, sevofluane at 3.5%.	5a. Rapid induction is desirable to reduce immediate stress. Stress is a critical factor in all procedures with guinea pigs.
6. Adjust oxygen flow to 3L/dl.	
7. When a suitable plane of anesthesia is achieved, remove the mask and swab the oral cavity, removing all debris from the cheeks and molars.	
8. Replace the mask, and proceed with induction and maintenance; reduce gas flow to 1.5%–2.0%.	
9. Carefully monitor heart and respiration rates.	
10. Frequently check the oral cavity, and remove any debris present. This can be done quickly by removing the mask, examining the oral cavity, and replacing the mask.	10a. This may be required several times during the anesthesia period to prevent aspiration. Each time should be recorded on the anesthesia monitoring sheet, along with vital signs and any other occurrence or drugs administered.

Surgical Preparation

OBJECTIVES

Upon completion of this unit, the reader should be able to:

▶ Prepare the patient for surgery

GUIDELINES FOR SURGICAL PREPARATION OF THE PATIENT

Guinea pigs should not be fasted for more than 2 hours prior to surgery. All patients should have a biochemistry profile prior to surgery, and it is recommended to perform another profile post-surgery to monitor metabolic changes.

Prior to induction and surgical preparation, follow all standard protocols using the aseptic technique. Review anesthesia procedures for anesthetizing the guinea pig.

UNIT ELEVEN

Radiology

OBJECTIVES

Upon completion of this unit, the reader should be able to:

▸ Restrain, position, and take radiographs in the non-anesthetized patient

GUIDELINES FOR RADIOLOGY PROCEDURES

Because of the inherent risk involved with general anesthesia of the guinea pig, patients usually are not anesthetized for routine diagnostic radiographs. The most common view taken is D/V rather than ventral/dorsal (V/D). Unless critically ill, a guinea pig will resist being positioned dorsally or laterally and will immediately try to right itself. The patient can be placed directly onto a film cassette that is large enough to accommodate the whole body. Because of the "freeze" behavior, the patient is likely to stay perfectly still for the radiograph. Though not ideal, this approach is often diagnostic. If a lateral view is diagnostically essential and the X-ray unit has a tube head that can be lowered and rotated, the patient can be placed in the same position as for a D/V, and the tube head can be lowered and rotated so the view taken is lateral.

PURPOSE

- To diagnose potential air pockets and areas of gut stasis
- To determine suspected deep abscesses
- To view the number and position of pups to be delivered by cesarean section
- To have an overall view of internal organs and potential fractures

COMPLICATIONS

- Positioning for a V/D or lateral view

EQUIPMENT

- High-density film and cassettes or digital plate
- Radiation safety equipment for personnel

PROCEDURE for Taking a Dorsal/Ventral Radiograph

TECHNICAL ACTION	RATIONALE/AMPLIFICATION
1. Have all equipment ready prior to bringing the patient into the radiology room.	1a. 3a. This prevents the patient from becoming alarmed and attempting to flee rather than freeze.
2. Place the patient directly onto the cassette or digital plate and quickly use the remote trigger to expose the film.	
3. If taking a lateral view, remove the patient from the room and reposition the tube head.	

FAST FACTS

Weight

- **Boar:** 900–1200 g
- **Sow:** 700–900 g

Life Span

- 4–7 years

Reproduction

- **Sexual Maturity**
  - **Boars:** 9–10 weeks
  - **Sows:** 4–6 weeks
- **Gestation:** 63 days
- **Litter Size:** 2–5 pups. Pups born precocial
- **Weaning Age:** 3 weeks (21 days)

Vaccinations

- **None**

Vital Statistics

- **Temperature:** 37.2°–39.6° C (99°–103.1° F)
- **Heart Rate:** 230–280/minute
- **Respiratory rate:** 42–100/minute

Dental

- Dental formula: 2 (1/1, C 0/0, PM1/1, M 3/3) = 20 teeth total

Zoonotic Potential

Bacterial:
- Salmonella
- Pasteurella spp
- Streptobacillus
- Staphylococcus

Viral:
- LCM (lymphocytic choriomeningitis): rare
- Rabies (potential—highly unlikely due to cavy exposure);

Parasites:
- Giardia spp (protozoan)
- Coccidian spp (protozoan)

Fungal:
- Dermatophytosis (ringworm)

Review Questions

1. What is the significance of drilling?

2. Describe the characteristics of a precocial neonate.

3. What is coprophagia, and why is it important to the newborn cavy?

4. What is the difference between rabbit pellets and guinea pig pellets?

5. Why must vitamin C be added to the daily diet of a guinea pig?

6. List five guinea pig foods that are high in vitamin C.

7. What disease is caused by a lack of vitamin C?

8. For what reasons should the water sipper tube be checked frequently?

9. Why should cedar shavings never be used as a substrate?

10. What is the reason that guinea pigs should not be housed with rabbits?

11. What is the main cause of peri-anal impaction in the guinea pig?

12. Excessive salivation can be cause by _____.

13. What are Kurloff bodies and where are they found?

14. Describe the anesthesia protocols specific to the guinea pig.

15. Why must a caesarian section sometimes be performed?

References

Judah, Vicki, & Nuttall, Kathy. *Exotic Animal Care Management* (2005). Clifton Park, NY: Cengage Learning.
Warren, Dean M. (2010). Small Animal Care & Management, 3rd ed. Clifton Park, NY: Cengage Learning.
http://www.adelaide.edu.au (accessed August 8, 2011)

CHAPTER 5

SMALL RODENTS

Animals are such agreeable
friends—they ask no questions,
they pass no criticisms.

— George Eliot

UNIT ONE

Overview of Species

OBJECTIVES

Upon completion of this unit, the reader should be able to:

▶ Compare and contrast the differences in small rodent pets, their general characteristics and behaviors

▶ Determine the sex of rodents

KEY TERMS

altricial	nocturnal
diurnal	pelts
estivate	sebaceous
fur slip	urethral cone
monogamous	Whitten effect

RODENTS

Small rodents are popular as pets because they can be housed just about anywhere. They are easy to feed and care for and often become a child's first experience in caring for a pet. Chinchillas, hamsters, gerbils, rats, and mice are all rodents. The Latin term *rodere* means *to gnaw*. They all have long, chiseling incisors that grow continually but are kept worn down by constant gnawing on any item available, and all rodents have developed strong jaw muscles for gnawing. Few things are impervious to gnawing rodents, and this must be carefully considered when housing a rodent. Rodents are coprophagic, so eating fecal pellets is normal and, for the young, essential in establishing normal gut flora.

CHINCHILLAS

The chinchilla is an appealing and extremely gentle rodent. Although they originally were bred commercially for their **pelts** (skin and fur), they have become popular pets because of their gentleness and appealing personalities. They usually are healthy and have a much longer life span than other rodents (14–15 years average). Chinchillas are small, compact animals, but their dense fur makes them appear bigger and rounder (**Figure 5-1**). Their fur is soft, and as many as 60 hairs grow from one hair follicle. They have a brush tail covered with hair that is only slightly coarser than the hair on their body. Chinchillas have four toes on each foot, with nail pads that never need trimming. They have long hind feet and are capable of jumping up and over things that are many times their own length. Though not strictly **nocturnal**, chinchillas are more active at dusk and during the night.

© Eric Isselee/Shutterstock.com

▲ **FIGURE 5-1** An adult chinchilla.

Behavior

Chinchillas have little means of defending themselves, but one mechanism that helps them escape capture is known as **fur slip**. If grabbed or frightened, they are able to release patches of fur. It may take 6 weeks or more for the fur to re-grow, so this is a major consideration in restraining the chinchilla.

Females can become cage territorial, especially when they have young kits. They stand upright on their hind legs, spitting and barking. If this fails to deter an intruder, they squirt urine with great accuracy at the perceived danger, be it another household pet or a person. Even if provoked to this point, they rarely bite.

▲ **FIGURE 5-2** A newly born chinchilla kit.

Breeding

Chinchillas are seasonally polyestrous, and the female's reproductive cycle usually is every 28–35 days. Pairs may be kept together for months or even years before they breed. Chinchillas are not nest-builders, and there may be no indication that the female is pregnant. Gestation is an average of 111 days. Parturition is brief and usually occurs in the morning. The average litter size is two, but a litter of three is not unusual, with one kit being much smaller than the other two and may not survive.

Chinchilla kits are precocial, and at birth they are fully furred with their eyes open and teeth erupted (**Figure 5-2**). They are active within minutes and nibble on solid food within a few hours after birth. Both parents tend to the young, and there is no need to remove the male; however, the female may be re-bred within 3 days postpartum.

Sexual maturity is variable and can be anywhere from 4–12 months. Both sexes have a **urethral cone**, which may be mistaken, in the female, for a penis. Male and female may appear similar at first, but the ano-genital distance, as in all rodents, is longer in the male (**Figure 5-3**). The male does not have a true scrotum; the testes are contained in paired sacs adjacent to the anus. The female has three pairs of mammary glands, but these are difficult to visualize unless she is nursing.

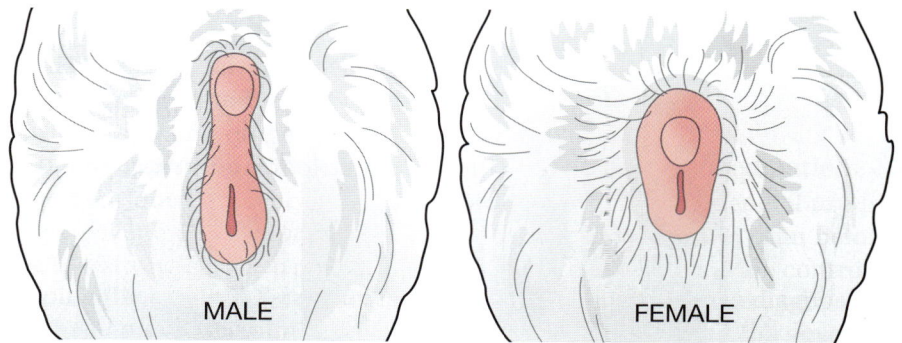

MALE   FEMALE

▲ **FIGURE 5-3** Comparison of genitalia in the male and female chinchilla.

Because of their dense fur, chinchillas need access to a dust bath to keep their fur clean. The dust should be about 1 inch deep and provided in a shallow bowl or pan. Chinchilla dust is a fine powdery pumice.

PROVIDING A DUST BATH

A good way to provide chinchillas with a dust bath is to use a one-gallon flat-sided fish bowl, placed in the cage flat side down. It provides ample room for their rolling antics, running in and out, and keeps the dust somewhat contained. It is not necessary to leave the dust in the cage at all times; providing it once or twice a week is sufficient.

Diet

Chinchillas are herbivores and should be fed a quality high-fiber diet consisting of free-choice timothy hay and formulated pellets. Pellets formulated for rabbits and guinea pigs usually are not as high in protein as those formulated for chinchillas. Another important factor is the length of chinchilla pellets. Chinchillas do not eat directly from a food bowl but, instead, pick up and hold individual food items. Chinchilla pellets are longer than other pellets so they are easier for the chinchillas to grasp.

HAMSTERS (GOLDEN OR SYRIAN HAMSTER)

Hamsters often are chosen as a first pet because of their small size and relatively low cost to house and feed (**Figure 5-4**). Problems quickly arise when owners realize they have a strictly nocturnal species, which does not willingly accept being disturbed during the day. Hamsters sleep deeply, and care must be taken to ensure that they are fully awake before attempting to pick them up to avoid being bitten. The chief complaint about the nocturnal habits of hamsters is that they make too much noise. It is not the hamster making the noise but, instead, the exercise wheel squeaking and tapping up against the side of the cage.

There is no method to train a nocturnal animal to meet the expectations of a **diurnal** pet and be awake during the day. Hamsters are not recommended as pets for young children because of their nocturnal habits and ability to quickly deliver a painful bite.

Behavior

As adults, hamsters are solitary. Attempting to keep two sexually mature animals together will result in fighting and the death of one of them. They should be put together only for brief, supervised periods to mate. The female is more aggressive and territorial than the male, and it is better if they are both placed in a neutral, supervised environment to breed.

▲ **FIGURE 5-4** A Golden hamster, also referred to as a Syrian hamster. This variety is called a Teddy Bear because of its long coat.

© iStockphoto/tunart

HAMSTER HOUSE-KEEPING ARRANGEMENTS

Hamsters are fastidious in maintaining and arranging their habitats. They choose specific areas for sleeping, stockpiling food, and depositing their waste. They do not rouse during the day to defecate but, instead, pick up fecal material from the sleeping den, place it in their cheek pouches and deposit the droppings in the specific area. Caregivers should be aware of the hamster's arrangement and not put items back in a different location after cleaning the cage. Hamsters become quite agitated and will start to rearrange the cage immediately. Hamsters are best kept at a comfortable room temperature. If the temperature exceeds 80° F, a hamster will **estivate**, or enter a dormant state, from which they are difficult to rouse.

Breeding

Male hamsters have large scent glands that appear as dark brown patches on either side of the back, near the hips. These are used for territorial marking. The testicles of a male hamster are evident and are often mistaken by owners as *suddenly appearing tumors* (**Figure 5-5**).

Females with a new litter should be left undisturbed, as they are known to cannibalize their young. Gestation is an average of 16 days. Hamsters are seasonally polyestrous, and most females do not have reproductive cycles during the winter months. The average litter size is four to twelve, and weaning usually is when they are 3–4 weeks old. Hamsters reach sexual maturity between 6–12 weeks.

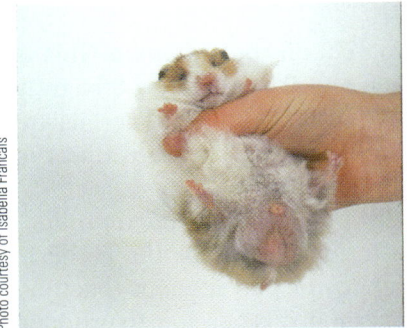

▲ **FIGURE 5-5** The testicles of a male hamster are clearly evident.

Photo courtesy of Isabella Francais

The average life span of a hamster is short, with an average of 18–24 months. It is difficult to approximate the age of a hamster when purchased unless it is acquired from a private breeder who has recorded the delivery date.

GERBILS

Though approximately the same size as a hamster, with an average body length of 5 inches, gerbils have haired tails that are as long as their bodies (**Figure 5-6**). Hamster tails are short and hairless. Unlike hamsters, gerbils do not have cheek pouches.

Photo courtesy of Isabella Francais

▲ **FIGURE 5-6** The Mongolian gerbil is a social and active small rodent.

Behavior

Gerbils are social with each other and need the companionship of other gerbils. Same-sex groups may be housed together if they are introduced at a young age. A pair (male and female) are **monogamous**; they mate for life and share in the care of their offspring. Owners should be aware of the social bonds that have formed and keep paired gerbils together. Gerbils are aware of their social community and protective of one another. If danger approaches, they thump with a hind foot to warn the others in the colony.

Gerbils are not strictly nocturnal and often are out and highly active during the day. They are playful with each other, scampering around the enclosure and stopping often for bouts of digging. They enjoy playing with small toys, and suggested items include cat balls with a small bell inside and cardboard tubes and boxes with several entrances to run through, around, over, and back again.

Breeding

Both sexes have an area of alopecia, a hairless patch on the abdomen. This is a large **sebaceous** gland used for scent marking, and it usually has a slightly orange pigment. Neither hamsters nor gerbils have sweat glands, but gerbils are able to tolerate slightly higher temperatures than hamsters, provided that the temperature is not accompanied by an increase in humidity.

Like hamsters, female gerbils are seasonally polyestrous. The reproductive cycle lasts approximately 4–6 days. The gestation period for the gerbil is 26 days—considerably longer than for the hamster. As the pair remains together, re-breeding occurs postpartum. During breeding season, it is possible to produce one litter per month, quickly leading to overpopulation and inbreeding. The average litter size is two to six, and the young are weaned completely by 4 weeks.

The young of hamsters and gerbils are born **altricial**, both blind and hairless. Gerbils are less likely than hamsters to cannibalize their young. The average life span for a gerbil is 3–5 years.

SERIOUS MEDICAL CONCERNS ARE RARE

Serious medical concerns are rare in the gerbil, but many gerbils are prone to seizure-like activity. These seizures are often triggered by the stress of handling or bouts of hyperactivity but rarely last longer than a few seconds. Owners should be cautioned to expect these incidents. The gerbil should be left alone and observed quietly during recovery. Attempts to assist in any way could result in a *latch-on* bite. Most gerbils recover without incident.

RATS AND MICE

Generally, rats are considered to be better pets than mice (**Figures 5-7** and **5-8**). Rats are easily socialized as pups and rarely bite. Male rats are more docile than females. Neutered male rats produce little odor. Rats are curious and intelligent and are capable of learning puzzle-solving skills, which they are able to retain and apply to new situations. They have been taught to retrieve items, play basketball with mini balls and a hoop and indulge in social play with their owners.

▲ **FIGURE 5-7** Rats can be very enjoyable small animal companions. There are several color variations. This rat, described by color, is a hooded rat.

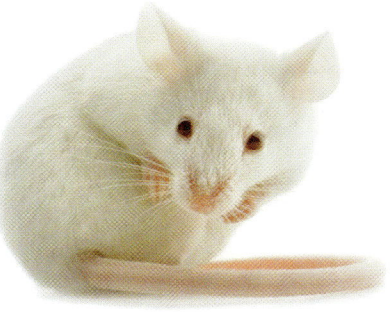

▲ **FIGURE 5-8** Mice can be very interesting to observe, but they are timid with people and quick to bite.

Behavior

Mice are more active than rats and rarely sit quietly long enough to allow much handling or interaction. They are much quicker to bite, with little provocation, and generally have a more nervous disposition. The odor of mice is much stronger than that of rats. Mice are less social with each other, often attacking or ganging-up on one cage mate. Rats rarely exhibit aggressive behaviors toward one another.

Breeding

Rats and mice both combine litters, care for, and nurse each other's young. Male rats are tolerant of pups and often hold them down and groom them. Male and female rats both pick up pups that have strayed and return them to the nest. Male mice tolerate pups to a lesser extent and frequently cannibalize pups found away from the nest.

Sexing rats and mice is similar. In pups, the sex is determined by the ano-genital distance, the length between the anus and the genitalia. In the female, the distance is much shorter than in the male. In the mature male rat or mouse, there is little doubt; the testicles are large and pendulous in lightly haired scrotal sacs.

Rats and mice reach sexual maturity between 6 and 8 weeks of age. Once mature, females will cycle every 4–5 days. Rats can be caged as a pair or set in small harems with one male and two or three females. In large colonies of mice, most females will not cycle until a new male is introduced.

THE WHITTEN EFFECT

When a different male is placed with a group of female mice, the females all begin to cycle at the same time. This is called the **Whitten effect**. Although it was first recognized in mice, synchronization of reproductive cycles occurs with many other species, including some agricultural species.

Average gestation time for mice is 16–18 days, with litters averaging eight to twelve pups. Gestation time for the rat is only slightly longer, an average of 20–22 days. Rat litters are often larger and can produce as many as 18 pups.

Diet

Rats and mice have survived for so long and in such numbers because they will eat anything. With companion rats and mice, however, the goal is not just survival but health and longevity. Commercial rodent blocks—those fed to laboratory animals—should also form the basis of companion rodent diets. They not only are nutritionally balanced but are also hard and require gnawing, which helps keep the incisors worn down. Various other rodent treats and seed mixtures can be offered, but those with an excessive amount of sunflower seeds should be avoided, as they are high in fat and small pieces of the shells sometimes become lodged in the oral cavity and tongue.

Restraint Techniques

OBJECTIVES

Upon completion of this unit, the reader should be able to:

▶ Correctly restrain the various species of rodents

KEY TERMS

exophthalmia

GUIDELINES FOR RESTRAINT

Each species requires a different method of restraint because each species is different, not only in their general behavior but also in their reactions to being handled, the perceived danger, and methods of escape. All rodents should be restrained gently but firmly and never startled by a quick move in an attempt to grab the animal.

PURPOSE

- To perform a health examination
- To determine the sex
- To perform a clinical procedure, surgery, or to take radiographs

COMPLICATIONS

- Fur slip (chinchilla)
- **Exophthalmia** (hamster)
- De-gloving of tail (gerbil)
- Bites to personnel

EQUIPMENT

- None. (Heavy gloves to prevent bites are not recommended as restraint tools for rodents. Gloves prevent a secure hold on the patient and the restrainer has little or no control of the amount of pressure being put into the restraint and the patient can be injured. Gloves increase the amount of resistance to restraint and the patient more easily escapes.)

PROCEDURE for Restraint of the Chinchilla

TECHNICAL ACTION	RATIONALE/AMPLIFICATION
1. Cover the examination table with a towel to make the surface less slippery.	
2. Place one hand under the abdomen and the let the chinchilla rest in the palm of the hand.	
3. The other hand should be wrapped around the base of the tail of the patient (**Figure 5-9**).	**3a.** This allows the chinchilla to be safely restrained if it decides to jump away. *Never* attempt to grab the body or scruff. The chinchilla will escape and you will be left only with a handful of fur and a chinchilla with a large bald spot (fur slip).
	3b. It is also appropriate to lift the chinchilla by the base of the tail and place the body in the palm of the other hand. This is a good technique if the patient is not used to being handled or is frightened by strange surroundings. *No attempt should ever be made to lift or restrain the patient by the ears!* This practice, as an excuse to prevent fur slip, is inhumane and causes damage to the fragile vessels of the pinna.

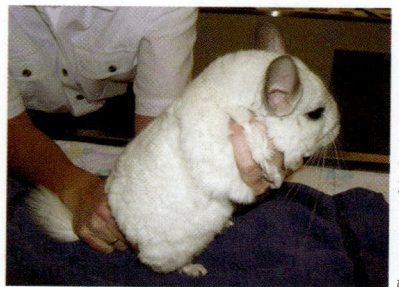

Photo courtesy of Jordan Applied Technology Center, West Jordan, Utah

▲ **FIGURE 5-9** Correct restraint of the chinchilla for an examination. One hand should be wrapped around the base of the tail to avoid the potential of fur slip.

PROCEDURE for Restraint of the Hamster

TECHNICAL ACTION	RATIONALE/AMPLIFICATION
1. Make sure the hamster is fully awake before attempting to reach in and pick it up.	1a. When first awakened, hamsters are easily startled. They stand upright or turn away, making hissing/chittering noises with their eyes still closed. This is *not* the time to reach in and attempt to pick it up. The result is likely to be a deep and painful bite.
2. When it is fully awake, gently prod it away from the sleeping site.	
3. Once the hamster is awake and moving, scoop it up in the hand.	
4. Allow the hamster a little time to settle into the hand, then restrain with a scruff.	4a. Always make sure the cheek pouches are empty before scruffing. Particles of food and bedding retained in the cheek pouches could cause injury.
	4b. All of the extra skin from the pouches must be included in the scruff so the hamster cannot turn around to bite.
	4c. Extended periods of scruffing may cause exophthalmosis; scruffing causes pressure behind the eyes, pushing them out of the orbits.
	4d. Always support the body of a scruffed hamster by cupping it in your hand (**Figure 5-10**).

Photo courtesy of Jordan Applied Technology Center, West Jordan, Utah

▲ **FIGURE 5-10** Hamsters can be scruffed with one hand and supported in the palm of the other hand.

PROCEDURE for Restraint of the Gerbil

TECHNICAL ACTION	RATIONALE/AMPLIFICATION
1. Simply pick up the gerbil and hold it in one hand or initially herd it into a cup (**Figure 5-11**).	1a. Gerbils are much quicker than hamsters, and scooping them into a cup gives the handler a better chance of keeping the gerbil confined before attempting a scruff.

Photo courtesy of Jordan Applied Technology Center, West Jordan, Utah

◄ **FIGURE 5-11** Gerbils can be held securely with one hand and may not have to be scruffed. They never should be lifted or held by the tail.

(Continues)

TECHNICAL ACTION	RATIONALE/AMPLIFICATION
2. When secured, use a gentle scruff and support the body with the other hand.	2a. Gerbils don't have the ability to "turn around in their skin" like a hamster because the skin is tighter and without the excess of cheek pouches.
3. *Never* attempt to catch or restrain a gerbil by the tail.	3a. One method of escape for the gerbil is that the entire skin of a trapped tail will de-glove (slide entirely off), leaving exposed tissue. The tail will have to be amputated surgically to prevent further trauma from the gerbil chewing on it and to prevent bacterial infection.

PROCEDURE for Restraint of the Rat

TECHNICAL ACTION	RATIONALE/AMPLIFICATION
1. Gently pick up the rat by placing one hand around the shoulders and support the body with the other hand (**Figure 5-12**).	1a. Refer to Figure 5-12. 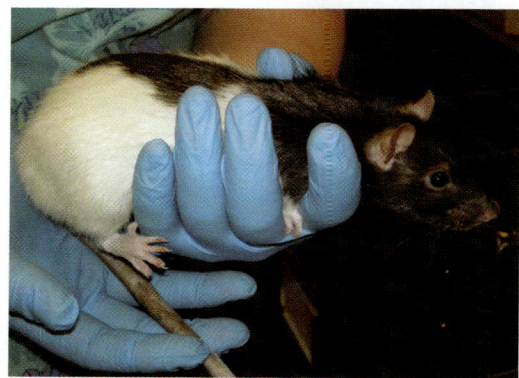 ▲ **FIGURE 5-12** Rats can be held securely with one hand around the thorax and shoulders.
2. If the procedure is likely to cause pain, place one hand around the shoulders as normal, but place the thumb under the mandible.	2a. This prevents the rat from biting in reaction to the pain.
3. *Never* pick up or carry a rat by its tail.	3a. Carrying a rat by the tail is not necessary and is one of the few instances that might provoke a bite during handling. From the rat's perspective, an object or predator has ahold of its tail.

PROCEDURE for Restraint of the Mouse

TECHNICAL ACTION	RATIONALE/AMPLIFICATION
1. Pick up the mouse by the base of the tail and, while maintaining the hold on the tail, quickly transfer it to a rough surface.	1a. When restrained, all mice will attempt to bite. If a mouse is held up by the tail for more than a quick transfer, it will quickly climb up its tail and bite.

(Continues)

TECHNICAL ACTION	RATIONALE/AMPLIFICATION
2. Be aware that the mouse will attempt to run away and stretch out away from the tail restraint.	
3. Use the other hand to collect it up in a scruff.	
4. Once securely scruffed, hold the tail with the little finger wrapped around it (**Figure 5-13**).	**4a.** Refer to Figure 5-13.

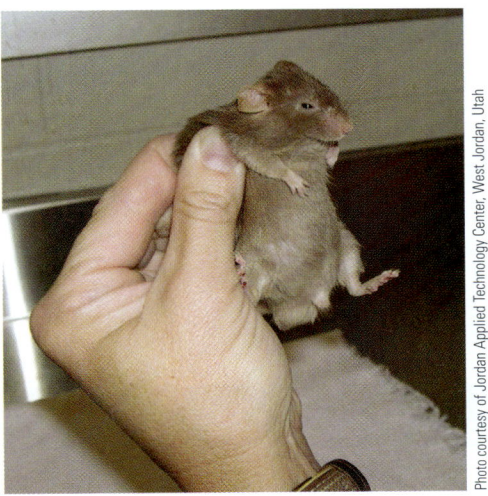

Photo courtesy of Jordan Applied Technology Center, West Jordan, Utah

▲ **FIGURE 5-13** Because of their small size and nervous disposition, mice should be scruffed for an examination.

UNIT THREE

Grooming

Upon completion of this unit, the reader should be able to:

▶ Perform grooming procedures for rodents

GUIDELINES FOR GROOMING RODENTS

Rodents, unless they are ill, are meticulous about grooming themselves. With rare exceptions, rodents do not need any extra grooming of their fur, their teeth, or trimming of their nails. If a rodent has excessively long incisors, it usually is because not enough material has been provided for gnawing. It is better to supply rodent blocks or hard wood for gnawing so the incisors can be worn down normally, rather than to attempt a tooth trim.

Chinchillas need a dust bath to keep their fur clean. However, male chinchillas can accumulate a tightly woven ring of fur around the urethral cone and penis. Eventually, as the hair builds up and the ring becomes tighter, the chinchilla will be unable to retract the penis. It becomes swollen because of the tourniquet effect of the hair ring, which cuts off blood supply and causes the penis to become necrotic. The fur ring must be removed to prevent tissue necrosis.

PURPOSE

- To remove a hair ring from the penis of a chinchilla

COMPLICATIONS

- None

EQUIPMENT

- Sterile lubricant
- Cotton-tipped swabs

PROCEDURE for Removing a Hair Ring on a Chinchilla

TECHNICAL ACTION	RATIONALE/AMPLIFICATION
1. Have an assistant restrain the chinchilla—one hand around the thorax, the other hand around the base of tail.	
2. Use cotton-tipped swabs to apply a liberal amount of sterile lubricant to the ring, penis, and urethral cone.	
3. Gently roll the hair ring up and off.	
4. *Never* use scissors in an attempt to cut away the hair ring.	4a. Attempting to use scissors could result in amputation.

UNIT FOUR

Patient History

OBJECTIVES

Upon completion of this unit, the reader should be able to:

▶ Complete a patient history on a rodent.

GUIDELINES FOR COMPLETING THE PATIENT HISTORY

Obtaining a patient history for rodents is the same as for other small mammals.

PURPOSE

- To obtain patient information from the owner to assist in a diagnosis
- To assist in a diagnosis and treatment plan
- To provide client education

COMPLICATIONS

- Incomplete information or history for the pet

EQUIPMENT

- Small-animal examination form
- Pen if recorded on hard copy

PROCEDURE for Taking a Patient History

TECHNICAL ACTION	RATIONALE/AMPLIFICATION
1. Complete small animal history form.	1a. Refer to history taking for rabbit or guinea pig.

Physical Examination

OBJECTIVES

Upon completion of this unit, the reader should be able to:

❭ Perform a physical examination

KEY TERMS

campylobacter	heaves
choke	nebulizer
dry heaves	otitis interna
dypsnea	porphryin
eczema	proliferative ileitis
Harderian gland	wet tail

GUIDELINES FOR PERFORMING A PHYSICAL EXAMINATION

The physical examination should follow as closely as possible the same system-by-system procedure as used for other patients. Because of the small size of some of the patients, it is not always possible to record heart and respiration rates or body temperature; however, it is always important to obtain an accurate weight in grams. The physical examination is often directed by the information obtained in the patient history, the chief complaint, and the technician's knowledge of the species and common problems associated with them. Only by knowing what is normal is it possible to recognize a medical problem or an abnormal behavior.

PURPOSE

- To obtain the most information to assist the veterinarian in a diagnosis and treatment plan

COMPLICATIONS

- Inexperience with various rodent restraint techniques
- Escape by the patient

EQUIPMENT

- Gram scale with a lidded bucket

PROCEDURE for Physical Examination of Chinchillas

TECHNICAL ACTION	RATIONALE/AMPLIFICATION
1. Evaluate the body posture.	1a. The normal body posture of chinchillas at rest is slightly hunched. Once they become interested in something, they often sit upright on their haunches. Although usually very calm, a chinchilla that becomes alarmed may stand upright on both hind legs. If greatly alarmed, either sex will urinate on the perceived threat. Allow the patient to calm down, and proceed slowly.
2. Look for any areas of alopecia.	2a. If there are areas of alopecia, it could indicate that the patient was handled roughly, caught in the cage furnishings, or attacked by another household pet. Refer to the patient history for the type of cage provided and other household pets.
3. Gently encourage the patient to move around in the cage, and assess locomotion.	3a. Chinchillas are quick and agile, but if there is an indication of abnormal gait, the hind limbs should be examined more closely. The chinchilla may have injured a hind limb, or it may have been caught in an exercise wheel (**Figure 5-14**).

Courtesy of Eric Klaphake

▲ **FIGURE 5-14** This chinchilla fractured its leg when it was caught in an exercise wheel.

(Continues)

TECHNICAL ACTION	RATIONALE/AMPLIFICATION
4. Examine any droppings present.	**4a.** Normal chinchilla droppings look like grains of rice and may be either dark brown or black, depending on the diet. An incorrect diet will cause either diarrhea or constipation. Constipation is seen more frequently and usually is the result of not enough fresh food and/or a combination of obesity and lack of exercise. Diarrhea that is dietary-related can also be caused by too much fresh food or sudden changes in diet. Another cause of diarrhea can be stress or excessive environmental temperature (chinchillas are most comfortable in temperatures between 65° F and 75° F).
5. Evaluate respiration rate.	**5a.** If there is any sign of respiratory distress or what appears to be repeated attempts to vomit, the veterinarian should be called in immediately. Chinchillas are susceptible to **choke**, which is an esophageal blockage. Signs of choke include retching, **dry heaves** (nonproductive vomiting), excessive salivation, and difficulty breathing. Choke, though esophageal in origin, can put pressure on the trachea and obstruct the airway. Choke can be caused by pieces of food or a hairball. Chinchillas, like many rodents, cannot vomit.
6. Correctly approach and restrain the patient to obtain weight in grams.	
7. While restrained, examine the eyes.	**7a.** Because of the dust bath, chinchillas can develop conjunctivitis from irritation caused by dust particles. The chinchilla may squint or keep one or both eyes closed, or there may be a sticky discharge that prevents the eye from opening. If conjunctiva is present, the eyes should also be examined for any surface abrasions.
8. Examine the urogenital area.	**8a. 9a.** Look for a hair ring around the urethral cone and penis. Confirm the sex (refer to Figure 5-3).
9. Confirm the sex.	
10. Examine the plantar surface of the hind feet.	**10a.** Chinchillas that are housed on wire with no solid floor area frequently develop pododermatitis (**Figure 5-15**).

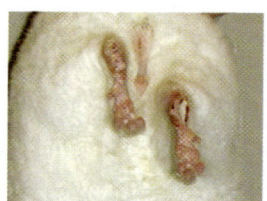

Photo courtesy of Jordan Applied Technology Center, West Jordan, Utah

▲ **FIGURE 5-15** Pododermatitis on the plantar surface of the feet.

11. Return patient to carrier or cage.

PROCEDURE for Physical Examination of Hamsters

TECHNICAL ACTION	RATIONALE/AMPLIFICATION
The physical examination for the hamster is much the same as that for the chinchilla. Areas of particular concern with the hamsters are the following:	
1. Make sure that the patient is fully awake before attempting to handle.	
2. Hamsters are highly susceptible to a condition called **wet tail**, so examine the rear quarters for signs of "slimy" wet fur, discharge, and an unpleasant odor.	2a. Wet tail is a **proliferative ileitis** caused by a **campylobacter** (bacterial) infection in the lower part of the small intestine, the ilium. Wet tail causes chronic wetness around the tail and quickly leads to dehydration and death. The condition can be brought on by stress, overcrowding, and poor husbandry practices. Even with prompt and aggressive treatment, the mortality rate is high.
3. Observe respiratory rate and note any signs of respiratory distress.	3a., 5a. Bacterial pneumonia is another common disease of hamsters. Signs include an ocular discharge and a purulent discharge from the nasal cavity. Prognosis is poor.
4. Scruff the patient, observe, and gently palpate any abnormalities (lumps).	4a. Various types of tumors are seen frequently in hamsters; they may be malignant or benign. Some tumors become so invasive that they are difficult to remove surgically, but all should be investigated (**Figure 5-16**).

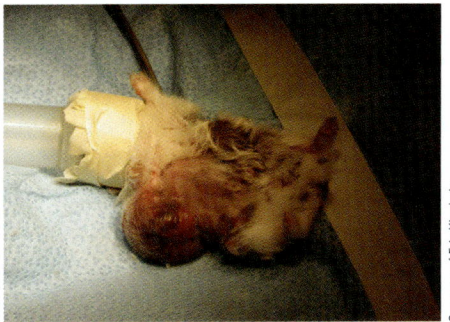

Courtesy of Eric Klaphake

▲ **FIGURE 5-16** Tumors are a frequent problem with hamsters. Surgical removal may be attempted, but with a tumor of this size, the survival rate is poor.

5. Observe the eyes and nose for signs of discharge.

PROCEDURE for Physical Examination of Gerbils

TECHNICAL ACTION	RATIONALE/AMPLIFICATION
1. Restrain patient correctly.	
2. Obtain an accurate weight in grams.	
3. Refer to patient history and chief complaint to complete the physical examination.	3a. Serious medical concerns are rare in the gerbil.

(Continues)

TECHNICAL ACTION	RATIONALE/AMPLIFICATION
	3b. *Red-nose* occurs in gerbils that are inappropriately housed in wire cages. Constant chewing on the cage bars irritates the end of the muzzle and nose, creating hair loss and inflammation. If not remedied, it may progress to facial **eczema** with lesions around the nose and moist dermatitis. Changing the habitat will improve the condition.

PROCEDURE for Physical Examination of Rats and Mice

TECHNICAL ACTION	RATIONALE/AMPLIFICATION
1. Follow steps 1–3 for gerbils.	**3a.** General signs of illness in rats and mice include weight loss, lethargy, and poor hair coat. They sit hunched and may exhibit piloerection. Erected hair is also a sign of pain in many animals, and greater care must be taken when handling them.
	3b. *Red tears* and *red sneezes* may be seen as ocular and/or nasal discharges. It is not blood but, rather, **porphryin**, a substance produced by the **Harderian gland**. This gland is present in most rodents and is located behind the eyes. Porphyrin is often evident in rats and mice that are overly stressed. Rats and mice that are ill still attempt to groom themselves, and the secreted porphryin can be transferred to other parts of the body, making them appear as if they are smeared with blood. (Rats in excellent health sneeze frequently, so this alone should not be interpreted as disease.)
	3c. Respiratory disease is seen frequently in rats and mice. In rats it usually is caused by bacteria, whereas in mice it just as likely is caused by a virus. Signs include **dyspnea** (difficulty breathing), loud and raspy respiratory sounds, weight loss, lethargy, hunched posture and, most indicative of all, **heaves** (abdominal muscle movement with each breath). The veterinarian may prescribe antibiotics, subcutaneous fluids, nutritional support, and bronchodilators. A **nebulizer** or vaporizer may also be used with an oxygen-infused isolation unit. (Cedar shavings should never be used as a substrate/bedding material for small mammals. When wet, cedar shavings produce toxic fumes that will cause respiratory problems.)
	3d. Rats and mice are both prone to **otitis interna**, a bacterial infection of the inner ear. Most present with a head tilt and reported behavior of constant circling. The direction of the circle usually indicates which ear is affected. This condition is treatable with antibiotic therapy.

(Continues)

TECHNICAL ACTION	RATIONALE/AMPLIFICATION

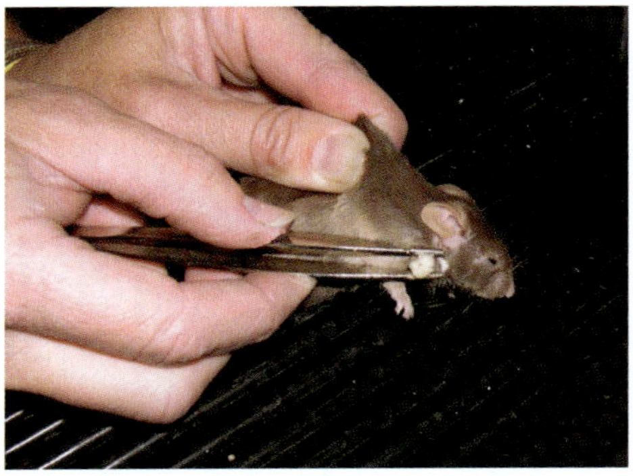

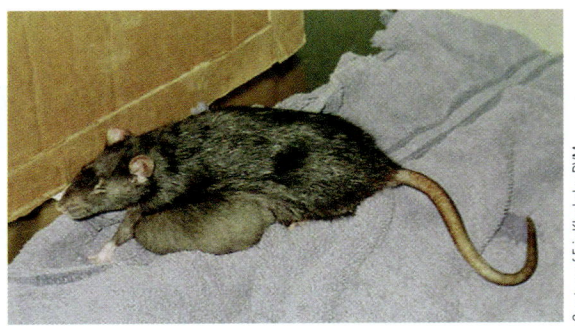

Photo courtesy of Jordan Applied Technology Center, West Jordan, Utah

▲ **FIGURE 5-17** The caseous material of an abscess does not drain, but should be approached with a small incision, and the contents removed. The site is allowed to heal as an open wound.

Courtesy of Eric Klaphake, DVM

▲ **FIGURE 5-18** An invasive mammary gland tumor on a rat, which could not be removed surgically because of the size and blood supply to the tumor.

3e. Abscesses occur frequently in both rats and mice. Abscesses are usually caused by an injury from some cage furnishing or by fighting. Abscess discharge in a rodent is thick and cream-colored, described as caseous, or cheese-like (**Figure 5-17**). Abscesses are opened surgically, the pus pocket removed then flushed with betadine, and allowed to heal as an open wound.

3f. Tumors are by far the most prevalent problem seen in rats and mice. Tumors grow quickly and have a rich blood supply. The only treatment is surgical removal, but unless they are removed when they are small, perioperative death can occur. This is attributed to the dramatic loss of blood and blood pressure related to the volume of circulating blood within the tumor (**Figure 5-18**).

UNIT SIX

Care of the Hospitalized Patient

OBJECTIVES

Upon completion of this unit, the reader should be able to:

▶ Perform the care required for the hospitalized patient, and prepare the patient for surgery

GUIDELINES FOR CARE OF THE HOSPITALIZED PATIENT

All rodents should be housed in escape-proof, chew-proof enclosures. Glass aquaria with screened and clamped lids are suitable, and they provide for easy access and observation of the patient (**Figure 5-19**). A sipper bottle should be provided for water. For food, heavy crocks are the most useful, as they are fairly chew-resistant and easily cleaned. For bedding, aspen shavings can be used, as well as plain newspaper or one of the commercially prepared paper litters. The cage should be kept out of the direct sunlight and maintained in ambient room temperature.

▲ **FIGURE 5-19** Suitable housing for the hospitalized rodent is a glass aquarium with a fitted screen lid.

UNIT SEVEN

Blood Collection

OBJECTIVES

Upon completion of this unit, the reader should be able to:

▶ perform a blood draw

GUIDELINES FOR BLOOD COLLECTION

Blood collection in small rodents can be difficult because the veins are small, fragile, and collapse easily. The best choice for blood collection in a rat or mouse is the lateral tail vein (**Figure 5-20**). This is not possible in the small, short tail of a hamster. In the gerbil, the tail is likely to de-glove unless the patient is anesthetized, and even then, a de-gloving may be the result. Blood samples from the hamster and the gerbil both may be obtained from the cranial vena cava. This procedure is not without risk, and the patient must be fully anesthetized. Because of the required restraint, the lateral saphenous vein is a good site to obtain a blood sample from the chinchilla.

When collecting blood from a rat or mouse, no more than 10% of total blood volume by weight should be taken. Blood volume from a healthy rat or mouse may be calculated from an average base volume of 70 ml/kg of body weight. Neither rats nor mice reach a body weight of 1 kg (2.2 pounds). Conversions must be calculated accurately with a calibrated gram scale to obtain the exact weight just prior to drawing blood. For chinchillas, the volume of blood drawn should not exceed 0.5 ml/100g.

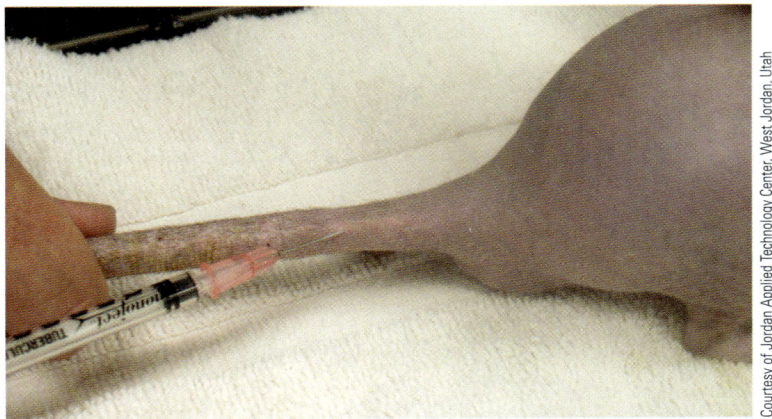

Courtesy of Jordan Applied Technology Center, West Jordan, Utah

▲ **FIGURE 5-20** Approach to the lateral tail vein in a rat or a mouse.

PURPOSE

- To obtain a diagnostic blood sample

COMPLICATIONS

- Very small veins
- Veins easily collapse

EQUIPMENT

- 25-gauge needle
- Tuberculin syringe
- Gram scale
- Calculator
- Empty syringe cases (3cc for mouse; modified 60cc for rat)
- Basin of warm water or warmed towel
- Microtainer tubes

PROCEDURE for Obtaining Blood Sample from a Rat or Mouse

TECHNICAL ACTION	RATIONALE/AMPLIFICATION
1. Manually restrain the patient or place head-first into an empty, modified 60cc syringe case for a rat or an empty 3cc syringe case for a mouse.	1a. If using an empty 60cc syringe case, the end of the syringe case has to be perforated from the inside (to avoid sharp points of plastic) or to have the tip cut off so the patient doesn't suffocate. Most 3cc syringes are already open-ended, but always check that this is the type you are using.
2. Once restrained, dip the tail in a basin of warm water or wrap it in a warmed cloth.	2a. The extra warmth will help dilate the veins for easier visualization and access.
3. With the tail fully extended and straight out from the body, insert a 25-gauge needle attached to a tuberculin syringe into the lateral tail vein (refer to Figure 5-20).	3a. A 25-gauge needle offers the best chance of avoiding vein laceration or collapse. A tuberculin syringe is an adequate size for the small amount of blood to be collected.
4. When blood flashes in the hub of the syringe, carefully withdraw the blood sample.	
5. Withdraw needle and apply pressure to the puncture site until bleeding stops.	
6. Transfer sample to a microtainer tube.	6a. This will preserve the sample and prevent it from clotting in the syringe.

Medication Administration

UNIT EIGHT

OBJECTIVES

Upon completion of this unit, the reader should be able to:

▶ Administer oral medications and injections

KEY TERMS

intraperitoneal

GUIDELINES FOR THE ADMINISTRATION OF MEDICATION

Whenever possible, oral medications are preferable to injections in these small rodents because of the pain and potential of tissue necrosis. Oral medications can be mixed with a variety of flavors and given with a syringe using the same method as for a cavy or a rabbit. Go slowly, allow the patient to swallow small increments, and avoid placing the medication in the cheek space. Oral medications should not be placed in the drinking water, as it is not possible to control the dose or to determine the effectiveness of the prescribed drug. Many rodents will avoid drinking medicated water and, aside from the other considerations, this has a detrimental effect on their hydration status.

If there is no alternative to oral medications, injections can be given subcutaneously (SQ), between the scapula or **intraperitoneal (IP)**. If administering an IP injection, the injection should be given in the lower right quadrant of the body to avoid the potential of injecting straight into or damaging an internal organ. Small amount of fluids may also be administered by these routes.

Anesthesia

OBJECTIVES

Upon completion of this unit, the reader should be able to:

▶ Induce, monitor, and recover small rodents during anesthesia

GUIDELINES FOR ANESTHESIA IN RODENTS

Most rodents are masked down with an inhalant anesthesia. A large canine mask is placed directly over the entire body of the patient so it sits flush with the table. Once a suitable plane of anesthesia is achieved—one without requirement of further restraint—the patient may be transferred to smaller ready-made mask placed directly over the nose and mouth. Eye lubricant or artificial tears should be placed in each eye to prevent oxygen and anesthesia gas from drying out the patient's eyes. A pulse oximeter can be applied to the foot or the tail to monitor heart rate and oxygen saturation (**Figure 5-21**). Small rodents are not fasted prior to anesthesia.

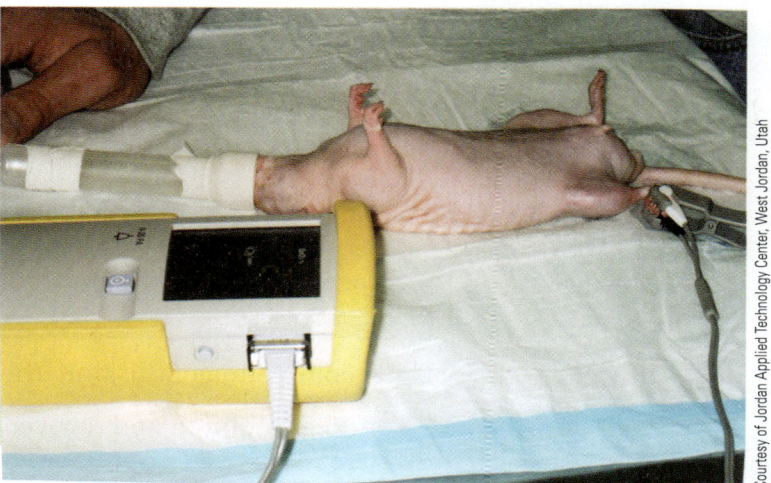

Courtesy of Jordan Applied Technology Center, West Jordan, Utah

▲ **FIGURE 5-21** Placement of pulse oximeter to monitor vital signs during anesthesia. The patient is a Hairless Rat, a variety originally bred for research.

PURPOSE

- To anesthetize patient for tumor removal
- To anesthetize patient for castration
- To anesthetize gerbil for tail amputation

COMPLICATIONS

- Difficulty in monitoring vital signs
- Surgical prep because of small size

EQUIPMENT

- Small-animal anesthesia machine with non-rebreathing system
- Large canine anesthesia mask
- Modified syringe cases
- Eye lubricant
- Pulse oximeter

PROCEDURE for Making a Small Rodent Mask

TECHNICAL ACTION

1. Depending on the size of the patient, use either a 20cc or a 35cc syringe case.

2. Remove the distal end (cut it off smoothly), and tape it directly to the end of the non-rebreathing system with an elbow fitting.

3. Form the mask by taping the palm part of a latex glove over the large end of the syringe case.

4. Cut a small slit in the glove, big enough to accommodate only the patient's head, yet not put pressure on the trachea (refer to Figure 5-21).

5. Trim the finger parts of the glove just below the taped glove for a neater finish.

UNIT TEN

Surgical Preparation

OBJECTIVES

Upon completion of this unit, the reader should be able to:

▶ Prepare a small rodent for surgery

GUIDELINES FOR SURGICAL PREPARATION OF THE PATIENT

Once the patient is anesthetized, the hair may be clipped and the standard procedure for aseptic technique followed. Great care must be used in using the electric clippers on these small patients. The skin is delicate and may easily be burned or cut with the clipper blade.

UNIT ELEVEN

Radiology

OBJECTIVES

Upon completion of this unit, the reader should be able to:

▶ Obtain radiographic films

GUIDELINES FOR RADIOGRAPHIC TECHNIQUE IN RODENTS

All films and cassettes should be high-density or digital plates. Technique charts will have to be modified because of their small size. Patients are normally anesthetized for radiographs.

FAST FACTS

Chinchillas

Weight

➤ 400–500 g

Life Span

➤ 15 years (average)

Reproduction

➤ **Sexual Maturity:** 4–12 months
➤ **Gestation:** 111 days
➤ **Litter Size:** 1–3 (2 average)
➤ **Weaning Age:** 6 weeks

Vital Statistics

➤ **Temperature:** 37°–38° C (98.6°–100.4° F)
➤ **Heart Rate:** 120–160 breaths/minute
➤ **Respiratory Rate:** 50–60 breaths/minute

Dental Formula

➤ 2/11 I, 0/0 C, 1/1 P, 3/3 M) = 20 teeth

Rat

Weight

➤ 250–450 g (males heavier)

Life Span

➤ 2.5–3 years—average (30–36 months)

Reproduction

➤ **Sexual maturity:** 8–12 weeks (2–3 months)
➤ **Gestation:** 21–23 days
➤ **Litter size:** 8–18
➤ **Weaning age:** 4–5 weeks

Vital Statistics

➤ **Temperature:** 37.7° C (99.8° F)
➤ **Heart rate:** 250–450 breaths/minute
➤ **Respiration rate:** 70–115 breaths/minute

Dental

> **Dental Formula**

Mouse

Reproduction

> **Sexual maturity:** 8–12 weeks (2–3 months)
> **Gestation:** 18–19 days (average)
> **Litter size:** 8–12
> **Weaning:** 4 weeks

Vital Statistics

> **Temperature:** 38.5°–40° C (100.4°–104° F)
> **Heart Rate:** 320–760 breaths/minute
> **Respiratory rate:** 60–220 breaths/minute

Dental

> **Dental formula:** 2 (1/1 I, 0/0 C, 0/0 P, 3/3 M) = 16 teeth

Hamster

Weight

> 40–70 g

Life Span

> 2–2.5 years (24–36 months)

Reproduction

> **Sexual Maturity:** 6–12 weeks
> **Gestation:** 15–18 days
> **Litter Size:** 4–12
> **Weaning Age:** 5–9 weeks

Vital Statistics

> **Temperature:** 37.6° C, 99.6° F
> **Heart rate:** 200–400 breaths/minute
> **Respiratory rate:** 40–70 breaths/minute

Dental

> **Dental formula:** 2 (I 1/1, C 0/0, PM 0/0, M 3/3)

Gerbil

Weight

> 50–131 g

Life Span

> 2–3 years (24–39 months)

Reproduction

- **Sexual Maturity:** 9–12 weeks
- **Gestation:** 23–26 days
- **Litter Size:** 3–8
- **Weaning Age:** 3–4 weeks (gerbils)

Vital Statistics

- **Temperature:** 37.6°C, 99.6°F
- **Heart rate:** 260–600 breaths/minute
- **Respiratory rate:** 85–160 breaths/minute

Dental

- **Dental formula:** 2 (I 1/1, C 0/0, PM 0/0, M 3/3) = 14 teeth

Vaccinations

- **None (for all rodents)**

Zoonotic Potential (all rodents)

- **Bacterial**
  - Salmonella
  - Mycobacterium
  - Streptobacillus
  - Pastuerella
  - Yersinia pestis
  - Staphylococcus
- **Viral**
  - LCM (lymphocytic choriomeningitis) especially hamsters
  - Rabies (potential in all mammals)
- **Yeast**
  - None reported
- **Fungal**
  - Dermatophytes (Ringworm) (**all rodents**)
- **Protozoan (all rodents)**
  - Giardia
  - Coccidia

Review Questions

1. Describe the correct method for chinchilla restraint.
2. How do restraint techniques differ between rats and mice?
3. What could happen if a scruff is maintained on a hamster for too long?
4. Why should gerbils not be caught or held by the tail?
5. In neonates, what is the difference between precocial and altricial?
6. Describe how to calculate the total volume of blood that can be safely drawn from a small mammal.

7. What is wet tail, the cause of this condition, and which species is affected?

8. Seizures are common in gerbils. What events could trigger seizure activity?

9. Which vein would be used for blood collection in a rat?

10. What is the Whitten effect?

11. Where is the Harderian gland located, and what does it produce?

12. Describe the problem unique to male chinchillas.

13. How is the sex of a rat or a mouse determined?

14. How does a chinchilla defend itself?

15. What are two general characteristics of rodents?

References

Judah, Vicki, & Nuttall, Kathy, *Exotic Animal Care & Management* (2005). Clifton Park, NY: Cengage Learning.
Warren, D.M. *Small Animal Care & Management*, 3rd ed. (2010). Clifton Park, NY: Cengage Learning.
www.http://iacuc.ucsd.edu (accessed November 16, 2012)

CHAPTER 6

BIRDS

A bird doesn't sing because it has
an answer, it sings because it
has a song.

—Chinese proverb

UNIT
ONE

Overview of Species

OBJECTIVES

Upon completion of this unit, the reader should be able to:

▶ Recognize different species of birds frequently seen in practice

▶ Explain the differences in feeding behavior and socialization of various species

▶ State a basic understanding of dietary needs

▶ Appreciate and respect the human–bird bond

KEY TERMS

avian

cognitive

hand-fed

passerines

psittacines

soft bills

ventriculus

AVIAN

Birds have become the third most popular companion species, after dogs and cats. Because of their popularity, a variety of birds are being presented more often to small-animal practices for routine care and medical concerns. Veterinary technicians who are inexperienced with birds may find **avian** patients more of a challenge initially because of their anatomical differences and unique behaviors; however, as skills and techniques are acquired, there is a tremendous reward in caring for avian patients. Clients expect a high degree of knowledge and skills from the veterinary team.

PSITTACINES

Companion birds represent some of the species generally referred to as parrots (**Figure 6-1**). These species include the Amazon, macaw, cockatoo, conure, African parrots and their smaller relatives, budgerigar (commonly called a parakeet), cockatiel, and lovebird. All of these species are **psittacines**, or hookbills, because of their powerful hooked upper mandible. A unique characteristic of psittacines is the structure of their feet: All psittacines have four toes—two that point forward and two that point backward. In addition to climbing and perching, they use their feet to pick up and hold food items and other objects of interest. These birds do not swallow food whole but instead break off pieces of food and crack open the hulls of seeds and shells of nuts to consume only the inner kernel.

© Aaron Amat/Shutterstock.com

▲ **FIGURE 6-1** Blue and Gold Macaws, pictured here, are psittacines and are among the largest of the popular companion birds.

Behavior

Almost all companion parrots have been **hand-fed**, taken from the parent birds when they are a few days old and fed a specific formula by a human. As a result, they have bonded with people as flock members. Generally, they are social and accustomed to interacting with people. They are best known for their **cognitive** and speaking abilities. They are highly intelligent and aware, possess the mental ability of a 3–5 year-old human child, and often have a vocabulary to match their intellect. While parrots mimic a variety of sounds and repeat words they hear, they also use words appropriately in response to a situation or to request a food item. Because of this ability, veterinary staff must be careful of the words and expletives they use.

Parrots are by no means domesticated, most being only one or two generations removed from the wild. It is important to understand and appreciate this when approaching and interacting with avian patients. They retain their instinctive behaviors, and one of those instincts is the ability and willingness to scream loudly for attention or if they become alarmed. The volume of their screams is so great that many technicians who work routinely with large parrots wear ear plugs to protect their own hearing.

PASSERINES

Another group of birds seen occasionally in practice consists of the **passerines**, or perching birds. Many of these are song birds, such as the canary (**Figure 6-2**) and a variety of finches. These birds are enjoyed because of their liveliness and color. The beaks of passerines are small and straight. They also have four toes, three of these digits pointing forward and one directed backward. They do not use their feet in the same way that psitticines do.

These small birds are often seed eaters, but they also consume a variety of fresh vegetables and greens. Unlike the psitticines, this group swallows whole seeds and the outer hull is ground away by the **ventriculus**, or gizzard. Some species within this group do not eat seed diets, but feed mainly on a variety of fruits and insects. This group is commonly referred to as **soft bills**. These small birds are not usually hand-tame and require a different approach for the veterinary technician with regard to capture, restraint, and handling.

Clients invest a great deal of money in purchasing their companion birds— maybe several hundred to several thousand dollars, depending upon the species. Birds are much loved family members, and clients expect veterinary technicians to understand the bond they have with their birds–no less than with a dog or cat— and to be professional and competent in all facets of the care entrusted to them.

▲ **FIGURE 6-2** Canaries are popular small passerines that are kept for their beautiful songs.

Photo by Isabelle Francais

Restraint Techniques

OBJECTIVES

Upon completion of this unit, the reader should be able to:

▶ Demonstrate correct capture and restraint techniques for large and small birds

GUIDELINES FOR RESTRAINT AND HANDLING OF THE AVIAN PATIENT

All doors and windows to the examination room must be closed, and other staff members should be made aware that a bird is in the examination room. Even birds with trimmed wings are capable of flight if they are frightened.

Almost all companion birds (psittacines) are hand-tame and willingly step up to an extended pair of fingers. Hookbills often use their beaks to grasp something prior to stepping onto it. This should not be interpreted as an attempt to bite. Jerking away the proffered hand will create immediate distrust. If the patient steps toward the extended hand, it is not likely to bite unless something frightens it or the person attempts to grab the bird. The owner should not hand the bird to the technician, nor should the technician attempt to take the bird from the owner, but should allow the bird to come out of its carrier or to be placed on the exam room table first. Birds are possessive/defensive of their owners and may attempt to bite the intruding hand or its owner. The bird should be asked to step up on the hand, either from the table or from its carrier. It then should be placed on a perch in the examination room and allowed to settle for a few moments.

An appropriate and useful perch is a gram scale. The scale should be turned on and re-set to 00.00 prior to placing the patient on the perch. With a perch already in place and a minimal amount of handling, an accurate weight in grams is easy to record (**Figure 6-3**).

If the patient is a passerine, it is best left in its cage for the initial examination. When capture is required, all perches, dishes, and toys should be removed from the cage first so the bird does not harm itself in an attempt to elude capture. Gram scales are used for weighing small birds with the use of the "birdie bucket," a round pot with ventilation holes and a secure lid. This is placed on top of the scale platform (**Figure 6-4**). All birds should be restrained from the back. Birds do not have a diaphragm, and any pressure on the sternum (breast bone) will

▲ **FIGURE 6-3** A perch attachment on a gram scale can be used to weigh a bird with minimal handling.

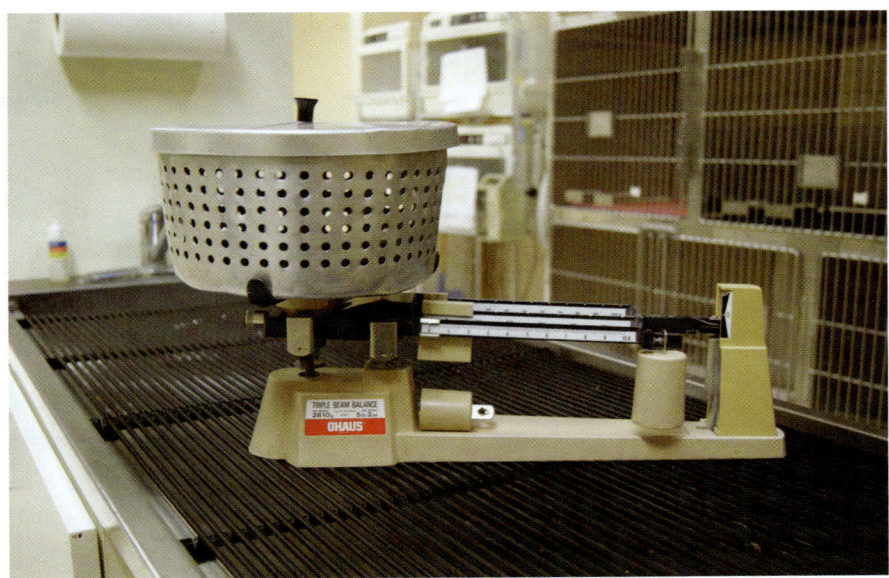

▲ **FIGURE 6-4** Small birds can be weighed with the use of a "birdie bucket" sitting on top of the gram scale.

compromise respiration. It is easier to capture birds for restraint when the bird is against the bars of the cage. Instead of attempting to chase the bird all around the cage, patience may be required.

A SMALL BIRD IS EASIER TO CATCH

A small bird is easier to catch in a darkened room. After all the cage furnishings have been removed for the bird's safety, have an assistant stand quietly by the light switch in the examination room. When the bird has settled, maintain a visual, then have the light turned off. With the sudden decrease in light, the patient is not likely to move.

PURPOSE

- To safely restrain the patient for a variety of clinical procedures
- To prevent injury to the owner, patient, and veterinary staff
- To cause the least amount of stress to the bird

COMPLICATIONS

- Owner interference
- Bite injury to owner or technician
- Escape from facility
- Flying into window glass
- Broken beak or feathers on wings or tail
- Bruising to facial area (especially in macaws)
- Regurgitation
- Fracture to wing or leg bones
- Highly stressed patient

- Respiratory compromise
- Death

EQUIPMENT
- Terry cloth towel for large birds
- Sheet of paper towel for small birds
- Gram scale with either perch or bucket in place

PROCEDURE for Restraint of the Avian Patient

TECHNICAL ACTION	RATIONALE/AMPLIFICATION
1. Have the towel draped over the hand, ready to capture.	1a. Many large birds are well aware of the "dreaded towel" and what it means. Dark-colored towels usually are less alarming than white towels. Do not wave the towel around, as it will further threaten the bird.
2. Quickly place the towel over the bird from the back, securing the head through the towel.	2a. Birds do not have a diaphragm. Respiration will be compromised if the restraint restricts movement of the sternum.
	2b. When the head is properly secured, the bird will not be able to bite.
3. When the head is secured, wrap the towel around the bird's body, securing the wings.	3a. This prevents the wings from flapping and often makes the bird feel more secure.
	3b. Check to be sure that the towel does not rub against the bird's eyes and there is no pressure applied to the cheek area. This is especially important in macaws because the area around the eyes is featherless and the skin is thin and easily bruised.
4. Use the free hand to uncover the bird's head, then grasp both feet.	4a. Birds will use their feet to grasp items being used in the procedure.
5. Hold the bird upright and close to the restrainer's body (**Figure 6-5**).	5a. Holding the patient upright is a more natural position for the bird and, less frightening for them. Holding the bird upright also reduces the possibility of regurgitation and aspiration of crop contents.

◀ **FIGURE 6-5** Large birds can be safely restrained with the use of a towel.

SAFELY CARRYING A BIRD ON THE HAND

A parrot is often carried on the hand to another area of the clinic for trims or other procedures prior to the towel restraint. In this case, with the bird perched on a hand, the thumb of that hand should always be placed over the front toes of the bird. This acts as a stabilizer and also prevents the bird from flying off. If the bird attempts to bite or fly off when carried in this manner, a quick, slight downward movement of the hand (maintaining the thumb in place) is effective. This easy correction, referred to as a "birdie earthquake," serves to remind the bird to sit quietly on the person's hand.

Birds should never be allowed to ride on a person's shoulder. In this position, it is impossible to be aware of the bird's behavior, and it also establishes a dominant position for the bird.

UNIT THREE

Grooming

OBJECTIVES

Upon completion of this unit, the reader should be able to:

▶ Perform routine trims of flight feathers, nails, and beak

KEY TERMS

blood feather	keratin
flashing	reminges

GROOMING

Courtesy of Ronie's For the Love of Birds

▲ **FIGURE 6-6** Equipment used to groom birds.

Avian patients are often presented for grooming procedures. Grooming includes trimming nails, clipping flight feathers, and trimming the beak. These procedures require two people; one is the restrainer, who also monitors the bird, and the other person performs the trims. **Figure 6-6** shows a variety of grooming tools used for bird trims.

PURPOSE OF NAIL TRIMS

- To prevent the nails from becoming overgrown to the extent that they inhibit normal movement and perching
- To reduce the chances of the bird becoming entangled in toys and other items in the cage
- To reduce sharpness so the owner may handle the bird more comfortably

COMPLICATIONS OF NAIL TRIMS

- Inadequate restraint
- Patient grabbing the equipment with a foot
- Incorrect equipment for the avian patient
- Breaking a toe
- Cutting into the nail quick
- Burning the plantar surface of the foot or other toes with the dremel-type tool
- Inadvertent amputation of digit, especially in small birds
- Overstressing the patient

EQUIPMENT FOR NAIL TRIMS

- Towel for restraint
- Motorized rotary tool (Dremel®)
- Two sizes of bird nail clippers
- Two sizes of human nail clippers
- Styptic powder to apply to nail end if cut too short

PROCEDURE for Performing Avian Nail Trims

TECHNICAL ACTION	RATIONALE/AMPLIFICATION
1. Have all equipment ready prior to restraining the patient.	1a. Efficiency of the procedure depends on having all equipment ready.
2. Have one person restrain the bird in the towel and the other perform the trims.	2a. The person restraining the patient should monitor the bird for any signs of stress.
	2b. In some species, especially the African Grey, stress causes small capillary leaks around the orbit of the cornea. Release the patient and allow it to recover before proceeding.

(Continues)

TECHNICAL ACTION	RATIONALE/AMPLIFICATION
3. Depending on the shape of the nail, a dremel-type tool is usually the first choice for a psittacine. If the nails are severely overgrown, cut them back using a nail clipper first. For small birds, a human nail clipper works best.	3a. Some nails may be so severely overgrown that they curl back into the foot. The toes should not be forced or twisted to achieve the trim, but first clip the nail just below the arc of the curve.
	3b. When using human nail clippers, hold them so the cutting blades are parallel to the nail and clipped from the side. This avoids splitting the nail.
4. Hold the foot between the thumb and the index finger; use the other hand to separate and trim each nail while protecting the other toes.	4a. Have the restrainer control the foot that is not being trimmed. It is important for the person performing the nail trims to protect the other three toes, especially when using the rotary tool.
5. Finish with a rotary tool or emery board for a rounded smooth tip.	5a. This removes sharp hooks that may be produced by the nail clipper.
6. If cut too short, apply styptic powder to stop the bleeding	6a. Styptic power should be used to stop the bleeding. Silver nitrate sticks are toxic to birds.

PURPOSE OF BEAK TRIMS

The purpose of most beak trims is to clean up the beak and remove the loose, uneven layers of **keratin**, the hard structure of epidermis that forms the beak and nails. Uneven layers of keratin can trap bacteria and accumulate debris from feeding and grooming. Occasionally, a therapeutic beak trim may be required if the beak is severely overgrown or deformed and has to be re-shaped. Before attempting a therapeutic beak trim, it is essential to know the normal beak shape for each species. Many clients ask to have the upper mandible trimmed because their birds bite. It is a mistaken belief that shortening the beak will reduce the pain of the bite. Doing so will actually cause harm to the parrot, as it will be less able to crack open seeds and nuts to remove the nutrient inside. The beak should be shortened only for therapeutic purposes when it is severely overgrown. The very tip of the mandible may be slightly rounded (touched-up) if it is "fish-hook" sharp. It is more important to understand why the bird bites its owner and to help resolve the behavioral problem.

COMPLICATIONS OF A BEAK TRIM

- Stress from restraint
- Potential seizure activity (especially in African Grey parrots): Procedure must be stopped immediately and the bird held safely and quietly to recover. Further use of the rotary tool is not recommended.
- Overgrooming of keratin layer, causing the beak to bleed. This undesirable result is not only painful but also may make the bird reluctant to eat. This is preventable with care and skill when performing the beak trim.

- Incorrect shaping of beak for function and by species.
- Feet inadequately restrained, allowing the patient to grab the rotary tool.

EQUIPMENT FOR BEAK TRIM

- Towel for restraint
- Motorized rotary tool
- Emery board
- Mineral oil

PROCEDURE for Performing Avian Beak Trims

TECHNICAL ACTION	RATIONALE/AMPLIFICATION
1. Have someone restrain the bird correctly so that the person performing the beak trims is able to grasp and control the bird's head.	1a. The restrainer must control the bird's feet and monitor the bird for signs of stress.
	1b. Birds are the only species with voluntary control of the pupils. **Flashing**—rapid constriction and dilation of the pupils—may result from alarm, excitement, or stress.
2. Trim the upper beak first (**Figure 6-7**). Use the rotary at moderate speed to lightly and quickly burnish the areas of the beak that need attention.	2a. This provides for absolute control of the head.
	2b. Care must be taken not to put pressure on the eyes or the cheeks.

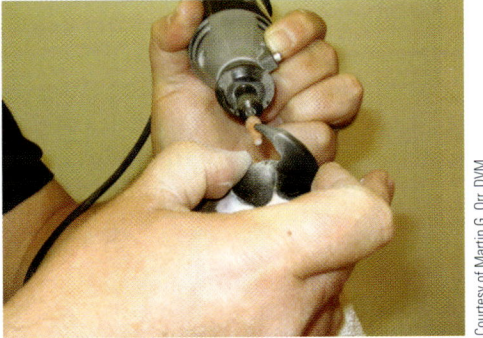

Courtesy of Martin G. Orr, DVM

▲ **FIGURE 6-7** Beak trims can be performed with the careful use of a motorized rotary tool.

3. Tuck the trimmed upper beak gently inside the lower beak to continue grooming the lower beak.	3a. This area usually requires the most attention because of the keratin layers.
	3b. The slower the speed of the rotary tool, the longer the procedure; however, the faster the speed, the easier it is to go "too far, too fast" and damage the beak.
4. Be aware of the beak structure: The lower beak may have slight vertical splits that could be worsened by trimming. Lightly buff the area below the crack with an emery board to create a *slight* horizontal notch.	4a. Gently place the upper beak in the foremost part of the lower beak to expose areas of concern.

(Continues)

TECHNICAL ACTION	RATIONALE/AMPLIFICATION
	4b. Proceeding with a dremel may widen the split and require a wire or epoxy repair.
	4c. Use the emery board to create a slight grove at the bottom of the crack to help stop it from progressing further.
5. Apply mineral oil to the trimmed beak.	**5a.** A light application of mineral oil removes the keratin dust created by the trim, gives the beak a nice clean shine, and pleases the client.

PURPOSE OF WING TRIMS

The purpose of trimming the primary flight feathers, **reminges**, is not to prevent a bird from flying but, instead, to limit its flying ability. Feathers should be trimmed in a manner that allows the bird to glide to the ground safely if attempting flight. Another reason the reminges are trimmed is for behavior modification. If an unruly bird learns and understands that it can escape from its owner by flying away to a place out of reach, this further complicates behavioral issues. Wing trims are also performed for the birds' safety; it prevents them from flying around the house and into window glass, into ceiling fans, onto a hot stove, or ending up head first in the toilet.

Clients are often quite particular about the style of wing trims they prefer and should be consulted. Correctly performed, a wing trim does not deter from the beauty of the bird. Clients sometimes request that the first two primary flight feathers of each wing be left untrimmed. This should be discouraged because these untrimmed feathers can easily be caught in the bars of the cage and broken off. Many clients like to keep the feathers from their bird trims, and it is not only polite but professional to ask them if they would like to have the trimmed feathers.

COMPLICATIONS OF A WING TRIM

- Cutting a **blood feather**, a new and growing feather with a rich blood supply
- Leaving a blood feather unprotected by trimming adjacent feathers
- Breaking the bones of the wing; caused by incorrect restraint and positioning
- Cutting the feathers too short so the cut ends become a source of irritation to the bird. This could lead to feather plucking and damage to the feather follicle.

EQUIPMENT FOR A WING TRIM

- Towel for restraint
- Small side-cutters or bandage scissors
- Hemostats or small pair of pliers, depending on the size of the patient

PROCEDURE for Performing Wing Trims

TECHNICAL ACTION	RATIONALE/AMPLIFICATION
1. Have restrainer hold the bird upright in a towel.	
2. Gently remove one wing from the confines of the towel and carefully extend it from the bird's body. Always keep the wing curved toward the body, never bent backward from the body. **Figure 6-8** illustrates the ventral view of a bird's wing.	2a. If the wing is bent backward away from the body, there is a great potential for fracture of the bones in the wing.

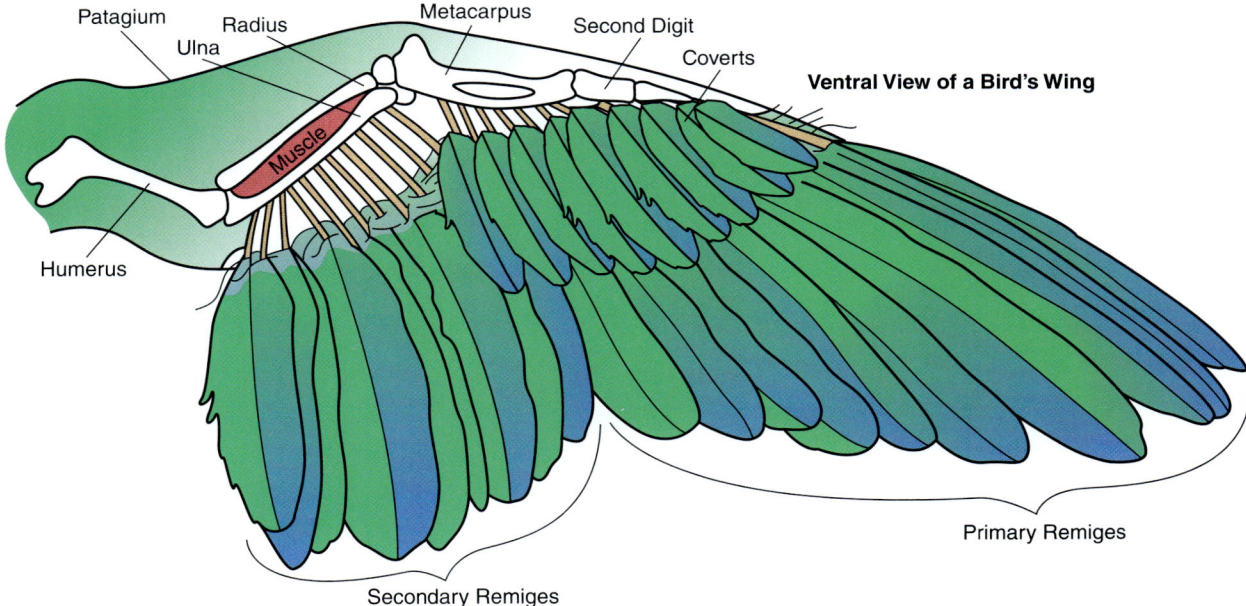

Labels on figure: Patagium, Ulna, Radius, Metacarpus, Second Digit, Coverts, **Ventral View of a Bird's Wing**, Muscle, Humerus, Primary Remiges, Secondary Remiges

▲ **FIGURE 6-8** Ventral view of bird's wing.

3. With the wing extended, isolate each feather prior to clipping.	3a. Check for feather health and feathers that may have been clipped, broken, or chewed already.
4. Check for blood feathers on either side of the feather that is to be clipped.	4a. With the wing extended, examine each feather from the underside, checking carefully for blood feathers. If a new feather is coming in, one feather on each side of the new one should not be clipped. This helps protect the new feather from damage.
	4b. Blood feathers are new and growing feathers containing a rich blood and nerve supply. A growing blood feather resembles an artist's paintbrush and, as it has not obtained full growth, it is shorter than the feathers adjacent to it. The shaft of a mature feather is clear and contains no blood.

(Continues)

TECHNICAL ACTION	RATIONALE/AMPLIFICATION
5. Begin with the outermost primary flight feather, and snip the feather shaft cleanly.	5a. The feather shaft should be cut through in one smooth action.
	5b. Never use scissors to cut across several feathers at one time because of the risk of cutting into a blood feather.
	5c. If a blood feather is clipped inadvertently, the entire feather shaft must be pulled. To do this, grasp the base of the feather firmly where it joins the body. Pull straight downward in the direction of the feather growth. This can be accomplished with hemostats, but for a large bird, a pair of pliers may be required. Apply pressure to the site until bleeding stops. Pulling a blood feather is painful, and anesthesia may be necessary, especially in large birds.
	5d. The veterinarian should be called for assistance to pull the feather and monitor the patient for stress and volume of blood loss. It is possible for a bird to bleed to death from a broken or cut blood feather, so immediate action must be taken.
6. Continue with each feather, checking carefully for other blood feathers.	
7. In heavier birds (Amazons, African Greys, macaws), it may be necessary to trim only the first five to seven primaries. In smaller birds (budgerigars, cockatiels), it may be necessary to trim more, if not all, the primary flight feathers.	
8. When finished, fold the wing back into its natural position to be held within the towel.	
9. Repeat the procedure with the other wing, and trim the same number of feathers.	9a. Trimming both sides evenly is necessary to keep the bird balanced. One wing should never be clipped differently from the other wing, or a severe trim performed on only one wing.
10. When trims have been completed, release the bird on the floor.	10a. Releasing the bird on the floor provides a period of adjustment to its new, limited flight status.

Patient History

OBJECTIVES

Upon completion of this unit, the reader should be able to:

▶ Complete an accurate patient history

KEY TERMS

dimorphic	molting
fomite	preening

PATIENT HISTORY

The veterinary technician should obtain a complete patient history and be able to perform a basic physical examination prior to the veterinarian seeing the client and the patient. Obtaining a complete history often provides important information that will assist the veterinarian in recommending changes in the environment and husbandry, determining which laboratory tests should be performed, and the general health status of the patient. Creating a medical history follows the same basic guidelines as that for a mammalian patient; however, there are some important differences with regard to obtaining a patient history for an avian patient.

PURPOSE

- To obtain as much information as possible to aid in a diagnosis
- To assist in assessing behavioral issues
- To help determine what tests may be required
- To evaluate the need for trims of beak, nails, and feathers.

COMPLICATIONS

- The bird may have come to the owner through a rescue organization and the prior history is unknown.
- The patient is presented as an emergency and/or in critical condition. If the patient is critical, no attempt should be made to examine the patient. Birds in critical condition should be placed in an incubator and monitored carefully until (or if) stable enough to proceed. Any handling of a critically ill bird could result in death.

EQUIPMENT

- Avian examination form or modified standard form
- Black ink pen, if the history is taken from the client away from the examination room/computerized client file.

PROCEDURE for Taking the Basic History

TECHNICAL ACTION	RATIONALE/AMPLIFICATION
1. Begin with the chief complaint, the reason for the visit.	1a. This could be a new-purchase exam, a new rescue, a request for grooming, or a medical concern.
2. Complete the signalment (species, age, sex) for the known information. If components are unknown, enter as "unknown."	2a. The technician should be familiar with the species. 2b. The exact age is not always known unless a hatch certificate is presented.

(Continues)

TECHNICAL ACTION	RATIONALE/AMPLIFICATION
	2c. Large parrots, with the exception of the Eclectus, are not sexually **dimorphic**—no visual differences between male and female. Positive sex determination of a psittacine is either through a laparoscopic examination of the reproductive tract or with a DNA sample. Because of the invasive nature of laparoscopy, the risk of surgery and anesthesia, sex determination generally has been replaced by submitting a feather sample for DNA testing. The only other reliable confirmation of sex is a history of egg-laying. Hens may lay eggs without the presence of a male. Chronic egg-laying is a problem frequently seen in cockatiels.
3. If the patient is a large bird and has arrived in a carrier, ask the client to describe, in detail, what type of cage the bird has, including inside dimensions, whether it has been painted and, if so, the type of paint used; kinds of food and water bowls (ceramic, plastic, etc.), and how they are placed within the cage; number, size, and types of perches; and location of cage within the house. What type and how many toys are in the cage? Is the cage near a window? Has the furniture in the room be re-arranged or new carpet installed? Basically, determine if there have been any recent changes to the room or location of the cage.	**3a.** The cage is an important aspect of the bird's habitat and husbandry. The cage should be large enough, *at a minimum*, for the bird to fully extend its wings in any direction without coming into contact with cage bars. Certain types of metal (e.g., zinc) and paint are toxic to birds. Food and water bowls should be placed within the cage so the bird doesn't defecate in them. There should be a variety in the size (diameter) and types of perches provided; perches should be non-toxic, unlikely to splinter (dowel rods or plastic), and easily cleaned. Birds should be provided with a variety of safe toys. Toys not only stimulate them but help a great deal to reduce boredom. Birds can be easily frightened when placed near a window. The heat from the sun is magnified through the glass. Birds are suspicious of change, and simply rearranging the room or hanging a new picture or mirror can cause anxiety. Some new carpets have been reported to be toxic to birds. If the cage is in the kitchen, there is a risk of exposure to toxic fumes from overheated non-stick pans, self-cleaning ovens, and even some burned microwave foods.
4. Find out whether the bird has cage mates or if there are other birds within the household. If so, record information including general health status, recent deaths, interaction with the patient.	**4a.** There may be competition and jealousy between birds, and other birds in the household may harbor undiagnosed pathogens or parasites.
5. Ask if the bird is allowed time out of the cage. If so, how closely is it supervised?	**5a.** Birds are extremely curious about their environment. If allowed unsupervised freedom, they can be highly destructive—chewing on furniture, playing with switches, and potentially getting into toxins. There are multiple risks of injury—flying into a window or a ceiling fan, falling into boiling water, or drowning in the toilet. Birds should never be allowed out of their cages unless they are closely supervised.

(Continues)

TECHNICAL ACTION	RATIONALE/AMPLIFICATION
6. Find out the type of household cleaners used in the vicinity of the cage and the products used to clean the cage, the food, and the water bowls.	6a. Birds are highly sensitive to household cleaning products, especially if aerosolized.
	6b. There are commercially made, bird-safe products for cage cleaning. Clients should be advised to never use abrasive cleaners on any part of the cage or equipment.
7. Determine the bird's normal diet, including any human foods offered, supplements given, how often the bird is fed, and the bird's appetite.	7a. Birds enjoy variety in their diets and sharing food morsels with their human flock. Some foods are toxic to birds and should not be fed: rhubarb and onions have been cited and avocado is known to be toxic to birds.
	7b. Birds should never be allowed to take food from a human mouth. Aside from the zoonotic potential, it sends the wrong message to the bird. A bird that bonds with its human flock might also see them as potential mates, and part of avian courtship is to regurgitate food to feed their chosen mate. As a general rule, birds being fed a healthy and varied diet do not need supplements.
8. Ask about any change in the bird's droppings.	8a. Visual examination of the bird's dropping can often indicate a disease process (Refer to **Table 6-1**).

TABLE 6-1 Evaluation of Bird Droppings.

DROPPING CONSISTENCY OR COLOR	CAUSES
Red dropping:	
Bright red	Fresh blood from the lower GI tract
Dark red	Old blood from the upper GI tract
"Tomato soup" or chocolate-appearing dropping	Suggests lead poisoning
Chartreuse green diarrhea	Suggests septicemia, true diarrhea, chlyamdiosis
Undigested seeds in the dropping	Suggests a digestive disorder
"Popcorn" looking stools	Suggests pancreatic insufficiency
Very large droppings in the morning	Suggests the bird is not defecating overnight
Polyuria (excessive urine)	Suggests fruits/veggies in the diet, medication, diabetes
Yellow or yellow-colored urates	Suggests a liver disorder

9. Find out if there other pets in the household.	9a. Other pets in the household can contribute to stress in the bird, especially if they are allowed to harass the bird or if the bird views them as rivals. Asking about children in the home should be approached only if required to elicit more history. Clients often become defensive when asked about children in the home; some people resent the question and react to it as an invasion of privacy.

(Continues)

TECHNICAL ACTION	RATIONALE/AMPLIFICATION
10. Ask if the patient has recently been **molting** (shedding old feathers that are replaced with new ones) and if the owner has noticed a change in the condition of the feathers.	10a. Feather molting is a normal process for birds, but it also can be a stressful time for them. The new feathers coming in should be bright and glossy. The bird may seem to spend an excessive amount of time **preening**, grooming the new feather growth. New owners often are alarmed when the bird sheds old feathers. Birds must be handled gently during a molt to avoid damaging the new and growing feathers.
11. Ask if the bird has been given the opportunity to bathe or shower.	11a. Bathing is a necessary part of self-grooming. Birds may have access to a shallow dish, or owners may mist them with a water bottle. Some birds are taken into the shower with their owners. Special shower perches are available to help avoid any contact with soaps or shampoos.
12. Ask the client to describe any noticeable change in behavior.	12a. Any change in behavior should be noted to rule out a medical cause.
13. Ask if the client has visited different pet stores or attended bird fairs and expos.	13a. There is a potential for **fomite** transmission of disease when bacteria and viruses are carried on the owner's clothes, hands, and hair.
	13b. Clients who visit pet stores and bird fairs should be advised to wash their hands thoroughly and change their clothes prior to handling their own birds. Human hair can also be a source of fomite transmission.

UNIT FIVE

Physical Examination

OBJECTIVES

Upon completion of this unit, the reader should be able to:

▶ Perform a physical examination

KEY TERMS

cere	keel
choana	stress bars
cloaca	vent

© Lee319/Shutterstock.com

PHYSICAL EXAMINATION

Courtesy of Eric Klaphake, DVM

▲ **FIGURE 6-9** A critically ill Amazon parrot.

The purpose of a physical examination is to determine the bird's general physical health and mental attitude prior to performing any further procedures or diagnostic tests. Any bird that is presented in a critical condition should not be examined; it should be placed in an incubator immediately. Even the slightest amount of handling may lead to the death of a critically ill patient. **Figure 6-9** shows an Amazon in a critical condition.

THE PHYSICAL EXAMINATION FOR AN AVIAN PATIENT

The physical examination for an avian patient varies slightly from other species because of its anatomical differences. The following are exceptions for the avian patient: The oral mucosa cannot be used for CRT (capillary refill time), skin tenting to assess hydration status is omitted, and TPR (temperature, pulse, and respiration) is omitted because these findings may not be reliable. In addition, there are no palpable lymph nodes.

PURPOSE

- To assess and record objective findings
- To complete the physical examination prior to consultation with the veterinarian
- To determine the presenting complaint

COMPLICATIONS

- Patient presented in critical condition
- Inadequate equipment to complete a thorough examination

EQUIPMENT

- Gram scale with either perch or containment basket
- Towel or paper towel for physical restraint
- Pediatric stethoscope
- Various-sized metal specula for examination of the oral cavity
- Microscope slides to collect fresh fecal sample
- Culturettes to collect samples for culture and sensitivity testing if required
- Indelible marking pen for labeling samples

PROCEDURE for Performing Basic Physical Examination

TECHNICAL ACTION	RATIONALE/AMPLIFICATION
1. Observe the patient from a distance, and note general attitude and alertness.	1a. Birds are prey for many species including other flock members. Masking any signs of illness is instinctive. In addition to attitude, pay attention to how the bird stands and moves around on the perch. It should grip fully with both feet.
	1b. The tail feathers should be observed, watching for an up-and-down movement, tail bobbing, which usually indicates a respiratory problem.
2. Record overall physical appearance including feathering.	2a. Feathers should lie flat and be uniform all over the body. They should be clean and bright in appearance.
	2b. Some species, especially cockatoos and African Greys normally produce feather dust.
	2c. Any feathers that appear abnormal in color or shape should be noted, including **stress bars**, horizontal dark lines marking the feathers. These can indicate an underlying illness or undue stress at molting.
3. Obtain an accurate weight in grams.	3a. Prior to placing the bird on the scale, the scale should be set to zero. If a containment bucket is used or a perch is attached, the scale should be re-set to zero to accommodate the extra weight.
4. Catch and restrain for closer inspection. Unless the patient is very small and can be held in one hand, have another staff member restrain the patient.	4a. The correct restraint is less stressful to the patient and allows the other person to examine the bird carefully.
5. Examine the bird's eyes, ear openings, **cere** (the firm fleshy structure at the juncture of the beak and face), the nares, and general beak condition.	5a. The eyes should be bright and clear, and both pupils the same size, even if flashing.
	5b. The ear openings normally are covered with a light layer of small feathers. Gently move the feathers aside with the thumb, and check for any discharge or odor.
	5c. The cere should be fairly firm and without apparent injuries or crustiness. If it appears crusty, this may be an indication of a mite infestation, especially if the patient is a budgerigar or a canary.
	5d. The nares (nostrils) should be clean, without odor, and no evidence of a discharge or foreign body. Occasionally a seed hull becomes lodged in the nare.
	5e. The beak should be hard and be the correct shape for the species. Check for any areas of softness, cracks, or deformity.
6. With the use of a mouth speculum, inspect the oral cavity, including the tongue and **choana** (slit-like opening on the roof of the mouth).	6a. The oral cavity is usually fairly dry. Note any amount of mucous, areas of white plaque or odor.

(Continues)

TECHNICAL ACTION	RATIONALE/AMPLIFICATION
	6b. The choana (on the roof of the mouth) should have small but visible fingerlike projections, **papillae**.
	6c. The tongue should be free of injury or areas of plaque.
7. Continue down the neck to gently palpate the crop, and observe the skin condition.	**7a.** The crop, if palpable, should be soft without any areas of hardness.
	7b. With the thumb, ruffle the feathers of the crop to look for signs of redness or necrotic tissue. The skin of birds is normally pale/white.
8. Evaluate and record the muscling over the **keel** (breast bone).	**8a.** The muscling over the keel bone should be palpated. In underweight birds or birds in generally poor condition, the keel bone is a prominent ridge with little muscling. In obese birds, the muscling on either side of the keel bone creates a small valley or groove. The extent of muscling should be symmetrical on both sides of the keel bone.
9. Carefully extend one wing, and examine both sides; repeat with the other wing.	**9a.** Each wing should be examined from both sides. Check for blood feathers, broken or chewed feathers, and feather cysts that appear at the point of feather growth.
10. Examine the **cloaca**, the common passage for feces, urine, and reproduction.	**10a.** The cloaca, also called the **vent**, should be clean. A bird that is reproductively mature will sometimes "wink" the vent upon examination.
11. Look closely at the condition of the scales on the legs and feet.	**11a.** Scales on the leg and feet should lie smoothly with no disruption or discoloration. If the beak has a crusty appearance, the scales of the legs and feet are also likely to be crusty.
	11b. Each nail should be checked for injury, noting any that may be missing or damaged.
12. Examine the plantar surface of both feet.	**12a.** The plantar surface also should have scales. These scales may be worn down or completely abraded because of the type of perch used. Looks for signs of redness, swelling, open sores, or an accumulation of feces, which indicate a bird kept in foul cage conditions.
13. Use a pediatric stethoscope to assess the heart and respiratory sounds.	**13a.** A pediatric stethoscope is used to assess the quality of the heartbeat rather than the heart rate. An accurate heart rate is difficult to obtain and is not reliable. In large birds, the heart rate may vary from 150 to 300 beats per minute. In small birds, the heart rate can be upward of 1,000 beats per minute and still be within normal range.
	13b. The stethoscope is used to evaluate respiratory sounds. It should be placed over the left side of the trachea and the cranial and caudal air sacs.
14. Record all findings, and allow the patient to relax on a perch or return to the cage.	

UNIT SIX

Care of the Hospitalized Patient

OBJECTIVES

Upon completion of this unit, the reader should be able to:

▶ Care for the hospitalized patient

KEY TERMS

chlamydophila

gavage tube

PBFD (psittacine beak and feather disease)

psittacosis

GUIDELINES FOR CARE OF THE HOSPITALIZED PATIENT

Caring for a hospitalized avian patient can be somewhat more of a challenge than for other companion animals. The first concern is housing. Standard dog and cat cages are suitable for larger birds, but smaller birds will be able to escape through the bars of the door. Plexiglass cages are a suitable alternative, provided that there is adequate ventilation. A modified perch should be provided, appropriately sized to the species, easy to clean, and placed near the bottom of the cage floor. It is natural for birds to attempt to reach a high perch, which gives them a greater feeling of safety, especially when they are ill. A bird that is ill should not expend extra energy in climbing to a higher perch. The floor of the cage should be lined with newspaper, but not the glossy-colored advertising inserts; when wet, the colors may bleed, obscuring important observations of the bird's droppings.

A water dish should be placed within easy reach. Clean, fresh water should always be available. Water dishes should be cleaned and refreshed twice daily or whenever soiled. Food dishes should also be placed within easy reach for the patient and the amount consumed should be closely monitored as well as the types of foods consumed. Seed eaters will leave the hull of the shell in the dish, and a casual glance might suggest that there is food available when, in fact, there is nothing but empty hulls. Birds are suspicious of new foods and the client should be asked to provide the usual diet whenever possible. Food should be offered free choice and available at all times. It may be necessary to use a feeding tube, or **gavage tube**, to administer nutrition and medications.

A patient that is non-critical is usually most comfortable at room temperatures. Birds are better at conserving body heat than they are at dispersing excessive heat. A bird that is in an overly heated environment will soon become stressed. Signs of heat stress in birds include panting and holding their wings laterally away from the body in an attempt to cool down. A bird in this condition should be moved immediately to a cooler environment. Critically ill birds usually are housed in a temperature-controlled incubator to help preserve body heat. The temperature should be monitored carefully and maintained between 85° F and 87° F.

Ideally, all avian patients should be housed in a separate room that blocks visual contact with dogs, cats, and ferrets and that reduces noise and activity. Wherever the location, the cage should be covered partially to provide an area where the bird may retreat and rest with less disturbance.

Any patient suspected of having **chlamydophila** or **PBFD (psittacine beak and feather disease)** should be kept in isolation. Chlamydophilia is a zoonotic bacterial disease. In humans the disease is called **psittacosis** to distinguish it from other Chlamydia bacteria that are passed from human to human. PBFD is caused by a virus and is not zoonotic. Both diseases can be transmitted by inhalation of dried fecal material, feather dust, contaminated clothing, feeding utensils, and by handling infected birds.

UNIT SEVEN

Blood Collection

OBJECTIVES

Upon completion of this unit, the reader should be able to:

▶ Perform an avian blood draw

KEY TERMS

apterium

BLOOD COLLECTION

Blood samples in the avian patient are collected more frequently from the right jugular vein. Most birds have a very small left jugular vein or no visible vein. The right jugular vein is superficial and easy to visualize. Prior to obtaining a blood sample, the crop should be empty to prevent regurgitation and possible aspiration. Occasionally, an air sac may sit on top of the vein. To avoid inadvertent puncture, the air sac should be gently pushed aside.

In large birds, blood may also be collected from the medial metatarsal vein of the leg. Collection from this site should be approached with caution because the vein is in close proximity to the medial artery. The artery can be easily lacerated if the bird is allowed to move. Accessing this site may require anesthesia.

Another potential site is the cutaneous ulnar vein, located on the medial surface of the wing. Because this vein is superficial and the skin is thin, the area can be easily bruised and hematomas formed.

It is not recommended to obtain a blood sample by clipping a toenail short enough to make it bleed. This is painful for the bird, and the injured toe is easily contaminated and open to infection. It also may cause permanent disfiguration of the digit.

Only small amounts of blood are required. In most patients, 0.5–1.0 ml of blood will be an adequate sample. Blood volume in birds is normally 10% of body weight. Prior to a blood draw, all patients should be weighed on a gram scale to determine the maximum amount of blood that can be taken safely.

PURPOSE

- To obtain a blood sample that is adequate in volume to perform various laboratory tests with the least amount of stress and trauma to the patient.

COMPLICATIONS

- Inadequate restraint
- Collapse of vein
- Puncture of air sac
- Insufficient volume for testing
- Obtaining too great a volume at one time
- Hematoma at site

EQUIPMENT

- Towel and assistant for restraint
- 25–27-gauge needle attached to a 1ml or tuberculin syringe
- Heparinized microhematocrit tube for sample transfer
- Sterile saline or water to dampen and separate feathers

PROCEDURE for a Jugular Blood Draw

TECHNICAL ACTION	RATIONALE/AMPLIFICATION
1. Unless the patient is a small bird and can be held in one hand, have an assistant restraint the bird.	1a. The correct restraint is essential to prevent laceration of a vein and to collect a blood sample quickly, with minimal stress to the patient.

(Continues)

TECHNICAL ACTION	RATIONALE/AMPLIFICATION
2. Ascertain that the crop is empty prior to the restraint positioning.	**2a.** The crop should be empty to prevent regurgitation and possible aspiration.
3. For a jugular blood draw, position the patient laterally on its left side, with the neck gently extended, the head of the bird held securely, and the right wing adjusted slightly caudally (**Figure 6-10**).	**3a.** The right jugular is the most prominent vein. Most species have a very small left jugular vein

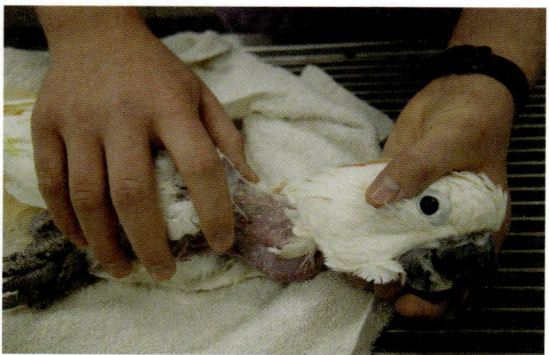

▲ **FIGURE 6-10** Restraint and positioning for a jugular blood draw.

TECHNICAL ACTION	RATIONALE/AMPLIFICATION
4. The restrainer is responsible for monitoring the patient.	**4a.** The restrainer's full attention should be on patient monitoring
	4b. The restrainer is responsible for applying direct pressure when the needle is withdrawn.
	4c. Coordinated teamwork can prevent hematomas and excessive bleeding from the site.
5. Expose the jugular by dampening the feathers with a sterile saline swab or water.	**5a.** Plucking feathers to obtain a blood sample should never be necessary. It is painful to the bird, as well as a source of irritation, which could lead to further feather plucking or self-mutilation.
	5b. Alcohol should not be used to dampen and separate the feathers. It has a cooling effect and contributes to delayed clotting time at the puncture site.
6. Move the feathers aside with the damp swab to expose an **apterium**, a featherless tract of skin.	**6a.** Throughout the feathering, there are tracts of featherless skin, apteria, making access easier.
7. If an air sac is sitting over the jugular vein, move it aside. Insert the needle caudally. When blood appears in the hub of the needle, withdraw the sample.	**7a.** If the air sac is inadvertently punctured, it will cause respiratory compromise.
	7b. Inserting the needle caudally provides a steady flow of blood and does not affect cranial venous pressure significantly.
8. Remove the needle and apply direct pressure to the site for a *minimum* of 30 seconds or until there is no further bleeding at the collection site. The patient should not be released until bleeding has stopped.	**8a.** In most patients, direct pressure will stop the blood oozing from the puncture site

(Continues)

TECHNICAL ACTION	RATIONALE/AMPLIFICATION
	8b. If the patient is released before bleeding stops, the bird will have to be caught and restrained again, causing greater stress, an increase in blood pressure, and more blood loss.
9. Transfer sample to heparanized microhematocrit tube.	**9a.** Heparin is an anticoagulant. The amount of heparin in a microhematocrit tube is correct for the small amount of blood drawn.

Medication Administration

OBJECTIVES

Upon completion of this unit, the reader should be able to:

▶ Administer an intramuscular injection

▶ Administer subcutaneous injections

ADMINISTERING MEDICATION

Prior to administering any substance, not only must the correct patient be confirmed, but also the drug or agent prescribed, dosage, route of administration, and treatment schedule.

INTRAMUSCULAR INJECTIONS

Intramuscular injections are given in the thickest part of the pectoral muscle, either side of the keel bone. If multiple injections are being administered, alternate sides should be used.

PURPOSE

- To effectively deliver a therapeutic drug with the minimal amount of and pain and stress.

COMPLICATIONS

- Inadequate restraint
- Regurgitation and aspiration
- Potential muscle necrosis at injection site
- Laceration of muscle tissue
- Inadvertent injection into vein

EQUIPMENT

- Towel for restraint
- Assistant to restrain patient
- Water-dampened cotton balls
- 25–27-gauge needle attached to tuberculin or 1 ml syringe
- Drug for injection

PROCEDURE for Administering an Intramuscular Injection

TECHNICAL ACTION	RATIONALE/AMPLIFICATION
1. Confirm that this is the correct patient.	1a. Giving any medication to the wrong patient is never acceptable.
	1b. If a drug is given to the wrong patient, it may have adverse effects on that patient and the patient requiring the medication does not receive the prescribed treatment.
2. Confirm the correct drug.	2a. Prior to administering any medication, the label should be checked when getting the drug from the supply cupboard, and re-checked to confirm the drug is correct prior to drawing it up into the syringe.
	2b. Administering the incorrect drug could potentially have dire consequences for the patient.

(Continues)

TECHNICAL ACTION	RATIONALE/AMPLIFICATION
	2c. Drugs, or any medication, drawn into a syringe should never be returned to the original vial. There is potential for contamination of the entire vial and the drug drawn is then wasted.
3. Confirm the correct dosage.	3a. Overdosing a patient with some medications could present complications including seizures and death. Under-medicating decreases the efficiency of the treatment and will have poor results.
4. Have the assistant restrain the patient.	4a. Correct restraint allows the needle to be held steady and avoids tissue damage.
5. Confirm that the crop is empty.	5a. Crop contents may be regurgitated with the risk of aspiration.
6. Separate the feathers over the muscle.	6a. Feathers should not be plucked over the injection site. Use a water soaked cotton ball to separate the feathers and expose the skin.
	6b. Plucking feathers over the injection site causes further skin irritation.
	6c. Using alcohol will delay clotting at the puncture site.
7. Insert the needle into the muscle mass.	7a. Inserting the needle should be done with one deliberate puncture.
8. Pull back the plunger to check for blood in the hub of the needle.	8a. Checking the hub of the needle confirms that a vein has not been punctured.
9. Quickly inject the drug into the muscle.	9a. By injecting quickly, there is less opportunity for the patient to struggle and disrupt needle placement.
10. Remove the needle and apply direct pressure to the puncture site.	10a. Apply direct pressure until there is no evidence of blood at the puncture site.

GAVAGE METHOD OF ADMINISTERING MEDICATION

The gavage method is used to administer oral medication directly into a bird's crop with the use of a gavage tube. The tube is attached to the syringe containing the medication. This method is used because birds do not accept medication by mouth. Adding medications to food or water is not a reliable method of ensuring that the patient is receiving the drug. Medicated food and water may taste and appear different and is not likely to be consumed. A patient refusing food and water will create additional problems of dehydration and malnutrition.

The gavage method is the most reliable, but it requires great care by the technician to prevent grave consequences if the procedure is not performed correctly. Birds also are given supplemental feedings with a gavage tube, and the oral

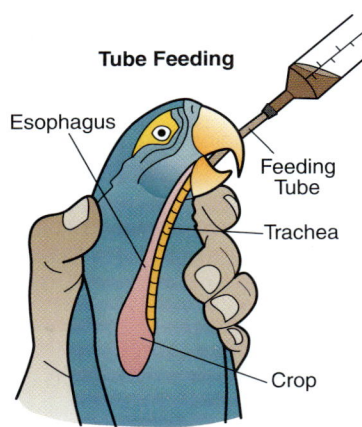

Tube Feeding

Esophagus

Feeding Tube

Trachea

Crop

▲ **FIGURE 6-11** Prior to administering medication or supplemental feedings by gavage tube, it is essential to confirm that the gavage tube has been placed in the crop and not in the trachea.

medication is added to the feeding formula **Figure 6-11** illustrates the correct gavage tube placement within the crop.

PURPOSE
- To deliver oral medications and/or supplemental feeding to the patient

COMPLICATIONS
- Rupture of the esophagus or crop
- Regurgitation
- Medication/food enters the respiratory system
- Aspiration
- Death

EQUIPMENT
- Towel and assistant for restraint
- Appropriate-sized gavage tube
- Syringe containing the medication and food formula

A RED RUBBER CATHETER MAY BE USED

A red rubber catheter also may be used, but because of its flexibility, it often is more difficult to place correctly. Even a small psittacine is capable of biting entirely through it. The metal tube is preferable because it cannot be bitten through, it has a ball tip that is easily palpated, and it can be passed without the use of a mouth speculum.

PROCEDURE for Administering Medication with a Gavage Tube

TECHNICAL ACTION	RATIONALE/AMPLIFICATION
1. The restrainer should hold the bird upright and facing the person who will pass the tube into the crop.	1a. Hold the patient upright to make tube placement easier.
	1b. The person placing the gavage tube has to have a clear visual of the tube progress into the crop.
2. Attach feeding tube to syringe prior to inserting the gavage tube.	2a. The syringe provides for greater control during the placement.
	2b. Attempting to attach the syringe after the tube has been placed causes greater movement of the tube and greater potential for damage to the esophagus and crop.
3. Gently insert the tube from the left side between the upper and lower beak.	3a. The esophagus and crop are on the right side. Inserting the gavage tube from left to right is less traumatic and access into the esophagus/crop is smoother than attempting to place the tube directly from the right side.

(Continues)

TECHNICAL ACTION	RATIONALE/AMPLIFICATION
	3b. In the left-to-right approach, pass the tip of the gavage tube diagonally across the glottis reducing the chance of inadvertent tracheal placement.
	3c. If the contents of the syringe are delivered through the trachea and enter the respiratory system, death can be immediate or within a few seconds.
4. Slowly and gently direct the gavage tube over the back of the tongue and down the right side into the esophagus and crop.	**4a.** The base of the tongue is a good landmark. The opening to the trachea also may be visualized if the patient is restrained correctly.
5. Palpate the progress of the feeding tube to confirm and *reconfirm* that the placement is correct.	**5a.** When the tube is correctly placed, the tip can be palpated while is it being progressed down the esophagus and into the crop.
6. When the ball tip of the feeding tube is confirmed to be in the crop, slowly infuse the contents of the syringe into the crop.	**6a.** The crop is a saccular pouch at the distal end of the esophagus. If the gavage tube has been correctly placed, the ball tip of the tube is readily palpated and if rotated gently, observed to be within the crop.
7. When the contents of the syringe have been delivered, slowly and carefully withdraw the feeding tube in the same manner only in reverse it will be from the right across to the left side.	**7a.** To prevent tissue trauma, as much care should be taken when withdrawing the gavage tube as when placing it.
8. Release the patient, and observe for several minutes for adverse reactions.	**8a.** Adverse reactions could include attempted regurgitation, complete regurgitation, yawning, and excessive head shaking.

FLUID THERAPY

Fluid therapy can be administered through intravenous or intraosseous catheters or by subcutaneous injection. Intravenous catheters are difficult to maintain in avian patients because birds chew through the delivery line or remove the catheter completely. If a large volume of fluid has to be delivered intravenously, the total volume is delivered in one injection. To keep the patient from moving and tearing the vein, anesthesia usually is required. Intraosseous catheters are placed more routinely than intravenous catheters because the veins are so fragile. Placement of an intraosseous catheter usually is performed by the veterinarian because it is an invasive (surgical) technique. This procedure requires anesthesia and surgical preparation of the site. The veterinary technician is responsible for anesthetizing the patient, surgical preparation of the site, anesthesia monitoring, and patient recovery. However, all technicians should be competent in administering subcutaneous fluid therapy.

PURPOSE

The purpose of fluid therapy is to restore and support normal hydration status.

COMPLICATIONS

- Inadequate restraint and positioning
- Fluid leaking from injection site
- Needle insertion too deep, and puncture of air sac
- Delivery of fluids into the air sac
- Patient death, in effect by drowning

EQUIPMENT

- Assistant for restraint
- 25–27-gauge needle
- Warmed lactated Ringers solution or 0.9% saline solution (as prescribed by the veterinarian)
- 3–6 ml syringe

PROCEDURE for Administering Subcutaneous Fluids

TECHNICAL ACTION	RATIONALE/AMPLIFICATION
1. Draw warmed fluids from the sterile bag into the syringe.	1a. Pre-warmed fluids prevent the patient from being chilled.
	1b. Birds that are ill have difficulty maintaining body heat.
2. Restrainer should hold the patient facing the person administering the fluids and completely controlling the opposite foot.	2a. If the patient is facing the person administering the fluids, it is easier to visualize the injection site.
	2b. If the other foot is not under control, the bird likely will attempt to grab the syringe, possibly dislodging the needle or driving it deeper in the tissue, or penetrating the air sac.
3. One leg is pulled slightly forward and lateral (**Figure 6-12**).	3a. The skin between the leg and the body has a greater subcutaneous space.

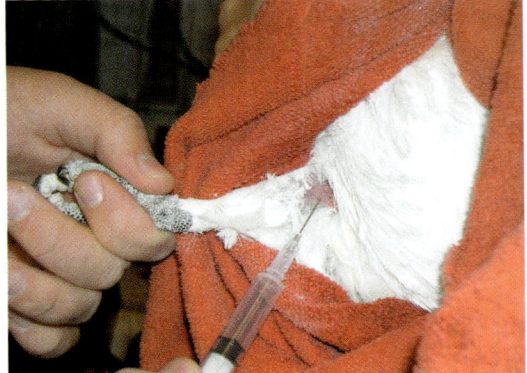

▲ **FIGURE 6-12** Subcutaneous fluids are delivered into the triangular area of skin between the body and the extended leg.

4. The needle is inserted into a triangle of skin formed between the leg and body.	4a. When the fluids are administered at a moderate rate, the bubble formed under the skin confirms that the fluids are in the correct location.

(Continues)

TECHNICAL ACTION	RATIONALE/AMPLIFICATION
	4b. A rapid injection of fluids could have dire consequences if the needle inadvertently punctures the air sac, in effect drowning the patient.
5. Inject the fluid bolus at a moderate rate.	
6. The fluids appear as a small bubble under the skin at the site.	

Anesthesia

OBJECTIVES

Upon completion of this unit, the reader should be able to:

▶ Deliver anesthesia and successfully recover the patient

▶ Place an endotracheal (ET) tube

GUIDELINES FOR AVIAN ANESTHESIA

Technicians should have a complete understanding of the anesthesia delivery system and be competent in delivering inhalant anesthesia, as well as monitoring patients and their recovery. There are many differences in avian anesthesia compared to other species. The physiological and anatomical differences must be considered when they are anesthetized, monitored, and during recovery. Air sacs function in respiration. Inhaled oxygen and anesthetic agents pass through the trachea and into the lungs to be further distributed to the air sacs. Birds do not have a diaphragm, and movement of the sternum must never be restricted.

Birds should not be fasted for more than 3 hours prior to anesthesia. They have a high metabolic rate and little glycogen-storage ability. Withholding food for longer periods could result in hypoglycemia. If the crop is full, it should be emptied prior to anesthesia to avoid regurgitation and possible aspiration.

Their unique respiratory system also makes birds prone to hypothermia. Heat loss occurs during anesthesia because of the constant need for air flow across the parabronchi. It is important to keep the bird warm while under anesthesia and to maintain positive pressure ventilation (PPV). Inhalent anesthesia is preferred for use in birds. Injectable anesthetics are rarely used because of the prolonged recovery time and possible muscle necrosis at the injection site. Advantages of an inhalant anesthetic are quick induction and recovery time with little cardio-pulmonary effects. Isoflurane and sevoflurane are both used successfully with a non-re-breathing system.

Vigilance is essential in assessing anesthesia parameters in the avian patient which differ from those normally used for a mammal. With induction, voluntary muscle control decreases but corneal and pedal reflexes remain. As the depth of anesthesia increases, respiration becomes more regular but vital signs continue to decrease and respiratory arrest may occur. Birds commonly experience periods of apnea during anesthesia. With close monitoring this can be prevented or reversed with PPV, and adjustment of the flow of anesthetic gas. If not corrected immediately, cardiac arrest is imminent and may not be reversible. Respiratory rate, rhythm, and depth of each breath are good indicators of anesthesia depth. It is essential to track the heart rate and pulse, either with a pediatric stethoscope or an EKG (cardiac) monitor. Palpebral and corneal reflexes are unreliable in the avian patient and papillary reflexes are not used because birds have voluntary control over the muscles of the iris. Lack of muscle tone in the wings and legs (no resistance when extremities are gently extended) are also good indicators that the patient has achieved a suitable plane of anesthesia.

Birds may be induced and maintained with the use of a small mammal face mask or the mask can be used for induction and subsequent intubation in larger birds. ET tubes should be uncuffed as the use of a cuffed ET tube may cause tracheal damage. It is not uncommon to experience mucous plugs in the ET tube during a prolonged procedure. If the patient is having difficulty breathing or the anesthetist is having difficulties ventilating the patient, the ET tube needs to be carefully checked for a mucous plug and immediately replaced with a new ET tube.

During recovery, birds should be wrapped in a towel and held upright until fully recovered. Extubation occurs only when the bird is alert enough to attempt to bite the ET tube. Keeping the bird intubated and upright will help prevent regurgitation. Most patients recover within five to ten minutes after the gas anesthesia is turned off. Because of their high metabolic rate, birds should be offered food shortly after recovery to prevent hypoglycemia.

AFRICAN GREY PARROTS

African Grey parrots usually have lower blood calcium than other species. Administering an appropriate dose of calcium gluconate by injection prior to anesthesia may help prevent seizures during anesthesia and recovery.

PURPOSE

General anesthesia is often used with avian patients for restraint and positioning in radiology, the elimination of pain during a medical procedure or surgery, and as a method of euthanasia. General anesthesia is less stressful for the avian patient than prolonged manual restraint or the use of restraint devices. Birds are induced smoothly and quickly, and their recovery is rapid.

COMPLICATIONS

- Hypothermia
- Bradycardia
- Uncorrected periods of apnea
- Induction with a full crop, subsequent regurgitation and aspiration
- Death

EQUIPMENT

- Towel for initial patient restraint
- Method of warming the patient—either circulating warm water pads or warm water-filled surgical gloves that are replaced frequently
- Predetermined dose of doxapram hydrochloride (Dopram) for respiratory emergency
- Immediate availability of warmed IV fluids and predetermined dose of epinephrine if cardiac emergency occurs
- Two appropriately sized, uncuffed ET tubes
- Appropriately sized small-animal mask for induction and/or maintenance
- Masking tape to secure the mask to the patient's head if required
- Gauze pads or cotton bandaging material to help prevent gas from escaping from collar of the mask
- Anesthesia delivery machine with a non-re-breathing system

PROCEDURE for Inducing and Maintaining Anesthesia in the Avian Patient

TECHNICAL ACTION	RATIONALE/AMPLIFICATION
1. The anesthetist uses towel restraint.	1a. The anesthetist usually can manage the restraint and placement of the face mask without an assistant unless the patient is very large and fractious.

(Continues)

TECHNICAL ACTION	RATIONALE/AMPLIFICATION

1b. The bird should be wrapped in the towel with the body tucked under one arm and the head pointing away from the restrainer's body.

1c. If the patient is a small bird, it can be held in the hand of the anesthetist.

2. Place mask over the patient's beak and nares (**Figure 6-13**).

2a. Gently situate the face mask over the beak and nares.

2b. If the patient is a small bird, the entire head may be placed in the mask.

2c. The mask may have to be modified for very small birds. Using the same modification for small rodents, tape the palm surface of the glove to cover the cone end of the face mask. Cut a small slit in the glove to accommodate the patient's head.

2d. Be aware that the bird will likely vocalize loudly, especially if it is a macaw. (Ear protection may be required.)

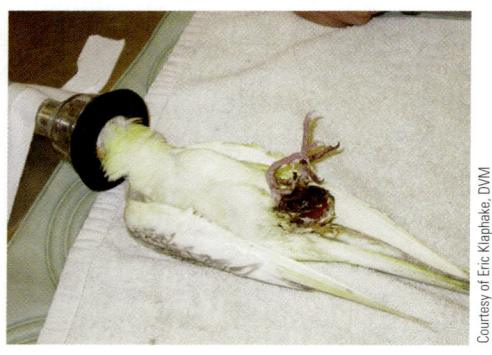

Courtesy of Eric Klaphake, DVM

▲ **FIGURE 6-13** Administering anesthesia to a small cockateil with a cloacal prolapse.

3. Open oxygen valve and allow the patient to breathe pure oxygen for a few seconds.

3a. Induction flow rate of oxygen should be between 1 to 3 liters/minute depending upon the size of the patient. It should never be below 1L/minute.

4. Introduce anesthetic gas.

4a. Isoflurane gas is most commonly used in avian patients. The induction rate is usually between 3% and 4%.

5. Monitor the patient carefully.

5a. Induction apnea is not uncommon in birds. The patient's respiration must be monitored carefully and the keel bone completely unrestricted.

5b. During any period of apnea, turn off the anesthetic gas and start PPV with pure oxygen until normal respiration returns.

6. Adjust flow rate of oxygen and anesthetic gas.

6a. Only with careful monitoring can the appropriate level of anesthesia depth be assessed and flow rates determined. On average, the isoflurane and oxygen rates can be reduced once the patient has become totally relaxed.

6b. *Generally*, birds can be maintained at an appropriate level of anesthetic depth by reducing the isoflurane concentration to 0.5%–2% and the oxygen flow at 1–2 liters/minute.

6c. Always lower the concentration of anesthetic gas before adjusting the oxygen. This helps prevent apnea.

6d. Slow, shallow breaths indicate that the patient is too deep, and corrective measures should be taken immediately.

(Continues)

TECHNICAL ACTION	RATIONALE/AMPLIFICATION
	6e. If respiration increases, it indicates that the patient is too light or the airway is occluded. If the patient is intubated, the ET tube may have a mucous plug. Be prepared to replace the ET tube immediately if an occlusion is present.
7. Continue vigilant monitoring for respiration and heart rate.	
8. Place patient in dorsal recumbency on a towel with a pre-warmed heating pad underneath.	
9. Proceed with intubation, or continue use of face mask for maintenance.	

INTUBATION OF THE AVIAN PATIENT

Small birds are rarely intubated because the necessarily small ET tubes frequently clog with mucus. In larger patients, intubation is common, especially when there is a facial or beak injury, which precludes the use of a mask. In severe cases, or an existing airway obstruction, the veterinarian may perform an air sac intubation.

PURPOSE
- To provide and maintain an airway during anesthesia
- To have immediate access for PPV
- To help prevent possible aspiration

COMPLICATIONS
- Abrasion of airway resulting from rough technique
- Incorrect size ET tubes or attempt to use a cuffed ET tube
- Substituting a red rubber catheter that the patient bites through
- Attempting intubation before the patient has reached a suitable plane anesthesia

EQUIPMENT
- Appropriately sized, uncuffed ET tubes (2)

PROCEDURE for Intubation

TECHNICAL ACTION	RATIONALE/AMPLIFICATION
1. Gently pull the tongue forward until the opening of the trachea is visible.	**1a.** Birds do not have an epiglottis, and the entrance to the airway is clearly visible.
	1b. The absence of the epiglottis leaves the entrance to the airway open. This is why there is such an increased risk of aspiration in avian patients.

(Continues)

TECHNICAL ACTION	RATIONALE/AMPLIFICATION
2. Insert the uncuffed ET tube, and connect to the anesthesia machine.	**2a.** Uncuffed tubes are used because the tracheal rings of birds are fragile and incomplete.

RECOVERING THE AVIAN PATIENT

Birds recover quickly once the anesthesia is turned off. Always turn off the gas first, and allow the bird to breathe pure oxygen for 1–2 minutes. Recovery from anesthesia usually is without complications. As birds begin to recover, they may vocalize, move the head, and attempt to flap the wings. They should be held upright, gently wrapped in a towel. Extubation occurs when the patient is alert but just prior to any attempt to bite the ET tube. Large parrots are quite capable of biting completely through an ET tube and may bite one in half, resulting in the necessity of recovering the aspirated portion. This can occur more easily with the use of softer, red rubber catheters.

Surgical Preparation

OBJECTIVES

Upon completion of this unit, the reader should be able to:

▶ Prepare the patient for surgery

SURGICAL PREPARATION OF THE AVIAN PATIENT

It is important to know what procedure will be performed, as well as the exact incision site. Unlike mammals, which are shaved for a larger sterile field, birds should have feathers plucked only from the incision site with the smallest margin possible and still maintain a sterile field. Even though most incisions are made through an apterium (featherless tract of skin) some feathers still must be removed. Always consult with the surgeon prior to plucking any feathers. When feathers are plucked, this should be done only when the patient is anesthetized and using a quick, downward motion, following the line of feather growth. Feathers should be plucked carefully so as not to tear the skin, and should be removed one at a time.

A modified surgical scrub is used, substituting sterile warmed water for alcohol to avoid chilling the patient. The feathers near the incision site can be moved aside and held away from the incision with the use of a sterile gel. When the incision is closed, the gel should be gently cleansed off the feathers. Birds normally preen, or clean their feathers, but in this instance, any remaining gel will quickly lead to the bird chewing out the sutures as a source of irritation. Often, a modified Elizabethan collar is used and placed on the patient prior to recovery. Birds with an E-collar should be observed carefully when they are fully recovered. When it is appropriately sized and placed, the bird should adapt reasonably quickly to the collar, be able to move and perch without tumbling, and be able to consume food and water without difficulties.

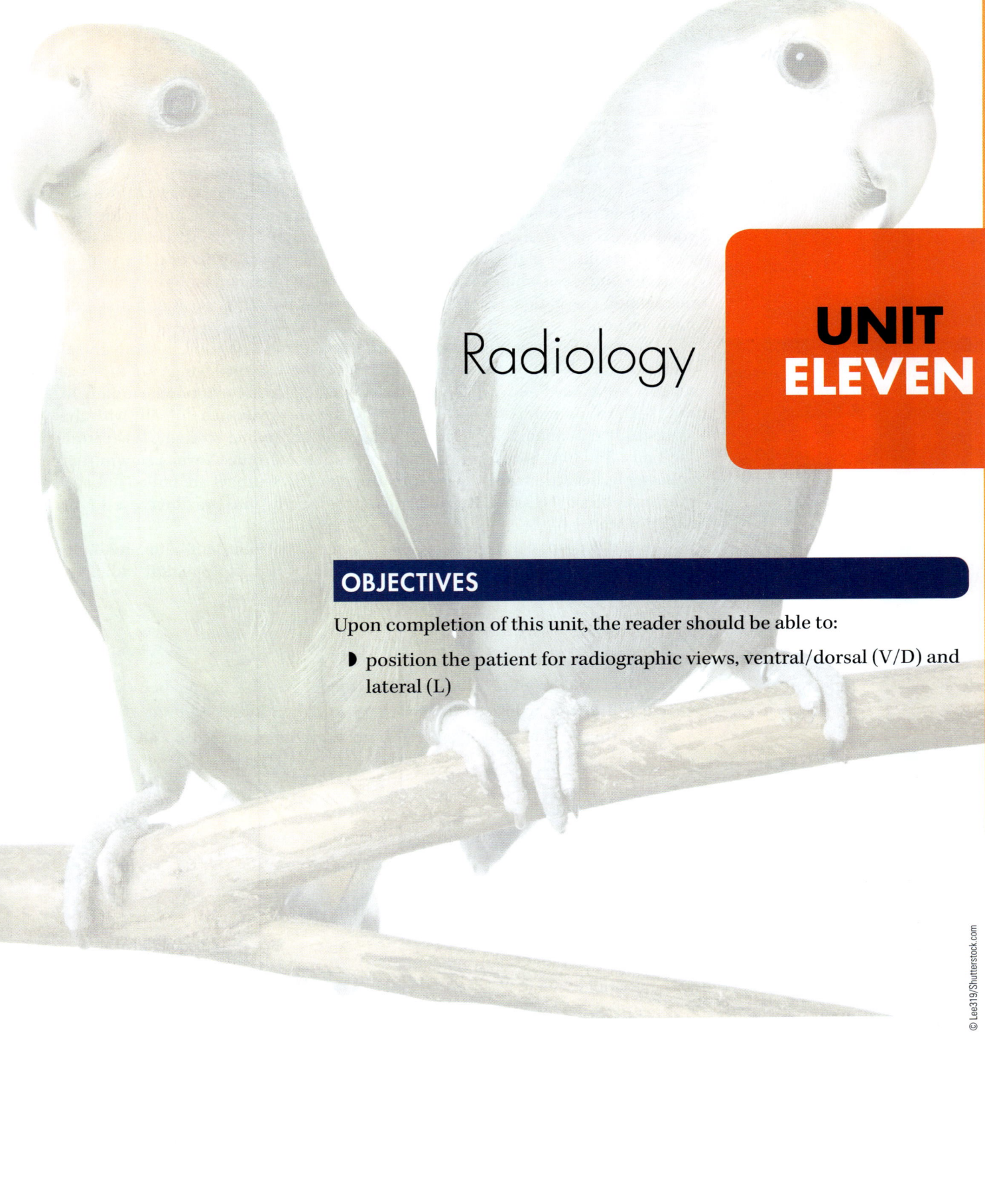

Radiology

OBJECTIVES

Upon completion of this unit, the reader should be able to:

▶ position the patient for radiographic views, ventral/dorsal (V/D) and lateral (L)

© Lee319/Shutterstock.com

GUIDELINES FOR RADIOLOGY

Avian patients are usually anesthetized prior to positioning for a radiograph. This not only prevents stress to the patient but also allows for better positioning without movement, removes the need for physical restraint, and avoids personnel proximity to scatter radiation. Prior to positioning, the patient must have reached a suitable plane of anesthesia. Fractures can easily occur if the patient struggles, and even when it is anesthetized, the technician must be careful when positioning the patient, to prevent fracture or dislocation. Positioning should never be forced.

The two radiographic views are lateral (**Figure 6-14**) and ventral/dorsal (V/D) (**Figure 6-15**). For a lateral radiograph, the patient's wings are fully extended and placed above the patient's back; the legs are extended caudally to expose the abdomen For a V/D view, the patient is positioned dorsally with the wings fully extended from each side. The legs are pulled straight down from the body. The bird is held in position with masking tape. Never use white or zonas tape, because when these tapes are removed, feathers are pulled out and there may be skin-tearing. Patients are taped directly onto the film cassettes or digital plate.

A standard small-animal X-ray machine is used. High-detail film and high-speed screen systems produce the best results if not using digital radiology plates.

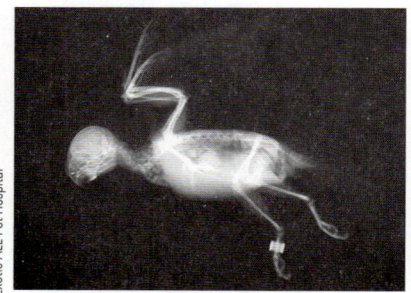

Courtesy of Martin G. Orr, DVM, Bird and Exotic ALL Pet Hospital

▲ **FIGURE 6-14** Correct positioning for a lateral radiographic image.

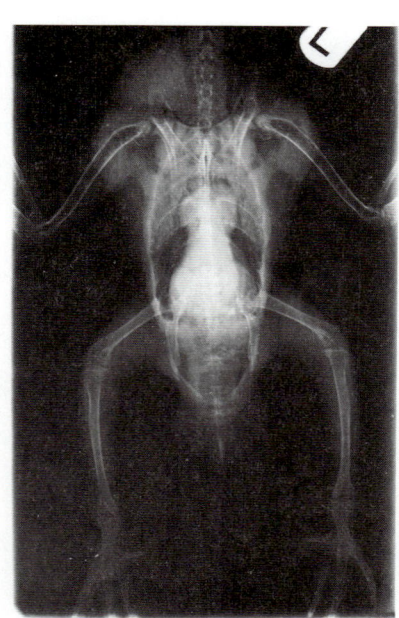

▲ **FIGURE 6-15** Correct positioning for a ventral/dorsal radiograph.

PURPOSE

- To obtain diagnostic information
- To determine the exact location of a fracture or possible foreign body
- To give direction to procedures that may need to be performed
- To determine underlying disease states or metabolic concerns

COMPLICATIONS

- Failure to monitor patient adequately during anesthesia
- Dirty film cassettes and resulting artifacts
- Incorrect positioning
- X-ray machine setting incorrect for species

EQUIPMENT

- Small-animal radiography machine
- High-detail film and fast-screen cassettes or digital plates
- Anesthesia machine
- Appropriately sized mask for patient
- Towel for induction restraint and recovery
- Roll of masking tape
- Lead apron, gloves, thyroid shields, and goggles
- Personal dosimetry badge
- Film marker
- Completed X-ray film identification imprint paper

TAKING an Avian Radiograph

TECHNICAL ACTION	RATIONALE/AMPLIFICATION
1. Prior to taking the patient to radiology, have all supplies and equipment ready.	1a. Everything required should be organized and ready for the patient.
	1b. Avian radiology should be a quick procedure. Being organized reduces anesthesia time and risk for the patient.
2. Confirm that it is the correct patient.	2a. When the correct patient is confirmed, write an X-ray identification tag and place it in the darkroom for imprinting on the film or place directly onto the digital plate.
3. Follow radiation safety.	3a. Everyone in the room should use radiation safety attire.
	3b. Because of the size of the patient, lead-lined gloves are difficult to manage. Use the foot pedal to snap the film and place your hands behind your back to prevent exposure.
4. Follow avian anesthesia protocol for induction and monitoring.	4a. The time under anesthesia is brief, but careful patient monitoring is essential.
5. Induce patient.	5a. When a suitable plane of anesthesia is achieved, there will be no resistance to manipulating the wings and legs for positioning.
	5b. Even though there is no resistance, the patient should be manipulated carefully to avoid fractures and/or dislocations.
6. Position on cassette for V/D view.	6a. To position for a V/D view, place the bird on its back with the wings fully extended, straight out from the body. The keel bone is used as a marker to make sure the view taken is correctly aligned.
7. Tape patient into position.	7a. Using masking tape, tape the wings in place. Extend both legs caudally and tape into place.
8. Remove anesthesia mask.	8a. Set the anesthesia mask to the side away from the cassette.
	8b. Leaving the mask on the patient may obstruct the view and a potential area of concern. In the short time required to snap the film, the bird will remain in a suitable plane of anesthesia.
9. Take radiograph.	9a. Step back from radiation table, and with your hands behind your back, use the foot pedal to trigger the exposure and take the film.
10. Return anesthesia mask to patient.	10a. Replacing the mask maintains a suitable plane of anesthesia while preparing for the second view.

(Continues)

TECHNICAL ACTION	RATIONALE/AMPLIFICATION
11. Remove tape, and reposition for lateral radiograph.	11a. Remove the masking tape gently, and discard. Pull it away in the direction of feather growth.
	11b. On the second cassette or digital plate, position the patient for the lateral view, confirming that there is no resistance.
	11c. The bird should be placed on its left side with both wings fully extended up and back. For consistency, always place the downside wing and leg cranial to the upside wing and leg.
12. Repeat steps 7, 8, and 9.	
13. Turn off anesthesia, then turn off oxygen.	13a. Remove the tape, turn off the anesthesia, and allow the patient to breathe pure oxygen for about 30 seconds. Turn off oxygen.
14. Recover the patient.	14a. Recover the patient fully before returning it to the cage.

IT IS IMPORTANT TO RECORD

It is important to record each setting of the radiograph machine, the views taken, and the species. This information can be used to develop an avian technique chart for future reference. Technique charts are guidelines only and depend upon the type of equipment available. It is also beneficial to take radiographs of deceased wild birds; for example, a sparrow is the approximate size of a canary, and a pigeon approximates an African Grey or a small conure. Modified small-animal techniques charts may be used; however, bone density and air sacs must be taken into consideration.

FAST FACTS

Finches

Weight
➤ 10–18 g

Life Span
➤ 4–5 years

Reproduction
➤ **Egg Incubation:** 18 days

Vital Statistics
➤ **Heart Rate:** 274 breaths/minute
➤ **Respiratory Rate:** 60–70 breaths/minute

Canaries

Weight
➤ 15–40 g

Life Span
➤ 8–10 years

Reproduction
➤ **Egg Incubation:** 18 days

Vital Statistics
➤ **Heart Rate:** 274 breaths/minute
➤ **Respiratory Rate:** 60–70 breaths/minute

Budgerigars

Weight
➤ 30–90 g

Life Span
➤ 6–8 years

Reproduction
➤ **Egg Incubation:** 18 days

Vital Statistics
➤ **Heart Rate:** 206–225 breaths/minute
➤ **Respiratory Rate:** 35–50 breaths/minute

Cockatiels

Weight
➤ 70–110 g

Life Span
➤ 18–20 years

Reproduction
➤ **Egg Incubation:** 21 days

Vital Statistics
➤ **Heart Rate:** 190–215 breaths/minute
➤ **Respiratory Rate:** 35–50 breaths/minute

Small Parrots

Weight
➤ 90–130 g

Life Span

➤ 20–30 years

Reproduction

➤ **Egg Incubation:** 22–24 days

Vital Statistics

➤ **Heart Rate:** 190–215 breaths/minute
➤ **Respiratory Rate:** 35–50 breaths/minute

African Grey Parrots

Weight

➤ 350–600 g

Life Span

➤ 40–50 years

Reproduction

➤ **Egg Incubation:** 24–28 days

Vital Statistics

➤ **Heart Rate:** 147–163 breaths/minute
➤ **Respiratory Rate:** 20–40 breaths/minute

Amazons

Weight

➤ 400–550 g

Life Span

➤ 70–80 years

Reproduction

➤ **Egg Incubation:** 26 days

Vital Statistics

➤ **Heart Rate:** 147–163 breaths/minute
➤ **Respiratory Rate:** 20–30 breaths/minute

Cockatoos

Weight

➤ 200–950 g

Life Span

➤ 30–40 years

Reproduction

➤ **Egg Incubation:** 24–29 days

Vital Statistics

➤ **Heart Rate:** 130–178 breaths/minute
➤ **Respiratory Rate:** 5–40 breaths/minute

Macaws

Weight

➤ 1000–1800 g

Life Span

➤ 50–60 years

Vital Statistics

➤ **Heart Rate:** 110–127 breaths/minute
➤ **Respiratory Rate:** 20–30 breaths/minute

Zoonotic Potential

➤ **Bacteria**
  ➤ C. psittaci
  ➤ Pasteurella
  ➤ Colibacilla
  ➤ Staphyloccus
➤ **Viral**
  ➤ Unknown
➤ **Fungal**
  ➤ Unknown
➤ **Protozoan**
  ➤ Giardia
  ➤ Coccidia

Review Questions

1. What are two distinct anatomical differences between psittacines and passerines, and how do these differences affect feeding behavior?

2. What major anatomical difference between birds and mammals determines the restraint technique?

3. Describe the correct restraint technique for a bird.

4. What are the basic grooming needs of a companion bird?

5. Describe three important differences in obtaining a patient history for a companion bird.

6. What are the exceptions in performing an avian physical exam that normally would be evaluated in a dog or a cat?

7. Describe the requirements in caring for a hospitalized bird.

8. List three potential blood draw sites and the complications of each.

9. Which site is used for intramuscular injections?

10. What is the gavage method and the greatest potential complication?

11. Which site is commonly used for the administration of subcutaneous fluids?

12. How do the air sacs influence anesthesia delivery?

13. What is a common problem encountered when an avian patient is intubated and what corrective action must be taken?

14. Describe surgical preparation of the avian patient.

15. What is the best combination of film and cassettes to use for avian radiology?

References

Ballard, Bonnie, & Rocket, Jody. *Restraint and Handling for Veterinary Technicians and Assistants* (2009). Clifton Park, NY: Cengage Learning.

Judah, Vicki, & Nuttall, Kathy. *Exotic Animal Care and Management* (2008) Clifton Park, NY: Cengage Learning.

McCurnin, Dennis M, & Bassert, Joanna M. *Clinical Textbook for Veterinary Technicians*, 5th ed. (2005). City: W.B. Saunders.

Sirius, Margi. *Principles and Practice of Veterinary Technology* 2nd ed. (2004). City: Mosby.

http://www.aav.org (accessed January 31, 2011)

CHAPTER 7

REPTILES

There is no greater diversity or
adaptability in form and function
than those to be discovered
among reptiles.

— *Author*

UNIT ONE

Overview of Species

OBJECTIVES

Upon completion of this unit, the reader should be able to:

▶ Identify common reptile species presented to the practice

KEY TERMS

chelonians

ecdysis

ectotherm

POTZ (preferred optimum temperature zone)

Salmonella

REPTILES

Reptiles have been kept in captivity for many years, but the knowledge to keep them healthy was severely lacking. There was no understanding of the importance of diet, lighting, temperature, and humidity requirements for various species. Most, if not all, were caught in the wild. Today, many reptiles are captive-bred and protected from collection in the wild. More information is available with regard to husbandry and dietary requirements, and reptile medicine has become specialized, with many advancements in diagnostics and procedures specific to their needs.

Owners have become more aware of the health issues with their reptiles, and more likely to seek veterinary advice and care instead of returning to the pet shop for advice or seeking information from an Internet chat room. In addition, reptiles are good at masking signs of illness, and by the time an owner becomes aware of a problem, the patient's condition may be critical.

Veterinary technicians have to recognize different species and to have a basic understanding of their husbandry requirements and specific behaviors. Also, veterinary technicians must be competent in restraint techniques, and be able to obtain samples, perform clinical procedures, and provide care for the hospitalized patient.

The species of reptiles seen as patients vary greatly, but most are lizards, snakes, and **chelonians** (turtles and tortoises). Lizards include iguanas—by far the most popularly kept reptile—geckos, chameleons, bearded dragons, Chinese water dragons, tegus, and monitors (**Figure 7-1**). Species of snakes that are frequently seen include pythons and boas, corn snakes, rat snakes, and water snakes. Like all reptiles, snakes vary in their temperament from placid to aggressive, depending on the species. Commonly seen turtles are semi-aquatic, and water turtles such as the red-eared slider and the soft-shell turtle. Popular tortoises are the African spur-thigh or sulcatta, star tortoises, and box turtles. In general terms, tortoises differ from turtles in that they are terrestrial vegetarians whereas turtles are aquatic, semi-aquatic, and omnivorous. Each group includes a variety of species, each with unique behaviors, husbandry requirements, diet, and temperament.

Reptiles are **ectotherms** (cold-blooded) and cannot generate or maintain their own body heat. They are dependent on the environment for body temperature, metabolic and reproductive health, and activity level. Body heat is acquired by basking in the sun, or in captivity, by providing heat and basking lamps. Each species has a specific environmental temperature that must be provided along with the appropriate amount of humidity. This is referred to as **POTZ**, the **preferred optimum temperature zone**. Table 7-1 lists the recommended average temperature ranges and humidity levels for some of the popular species.

Most health problems with reptiles are directly related to not providing the correct environment. In addition to POTZ, full-spectrum bulbs are required to provide ultraviolet light. Ultraviolet light assists in the assimilation of vitamin D3, which is necessary for the dietary absorption of calcium.

All reptiles have scales. Scales protect the skin and are shed and replaced along with the outermost layer of skin on a regular basis. This period of shedding, or **ecdysis**, is a stressful time for reptiles. Many refuse to eat, and if the POTZ range is incorrect, they may have difficulty in completing the shed.

Reptiles are carriers of **Salmonella**, a zoonotic bacteria. Precautions always should be taken to avoid contamination, especially when handling a reptile, its feces or soiled cage, and water and food dishes. Gloves should always be worn and hands washed frequently, particularly between handling different patients. All clients should be advised of the risk of Salmonella from their reptiles.

© Brooke Whatnall/Shutterstock.com

▲ **FIGURE 7-1** A variety of reptiles are seen in the veterinary practice.

TABLE 7-1 Reptile Temperatures and Humidity Levels

LIZARDS	POTZ	HUMIDITY
African fat tail gecko	78–90 °F	20–50%
Anole	73–84 °F	70–80%
Basilisk	73–86 °F	80–100%
Bearded dragon	84–120 °F	30–40%
Blue-tongued skink	81–90 °F	30–60%
Crested gecko	74–83 °F	50–80%
Dwarf chameleon	70–85 °F	70–80%
Giant day gecko	80–86 °F	50–80%
Fischer's chameleon	72–85 °F	70–80%
Gold dust gecko	80–86 °F	50–70%
Green Iguana	85–103 °F	80–100%
Jackson's chameleon	70–80 °F	50–75%
Leopard gecko	77–86 °F	20–30%
Panther chameleon	85–90 °F	70–100%
Savanna monitor	80–95 °F	20–30%
Tegu	78–90 °F	50–80%
Veiled chameleon	80–95 °F	20–30%
Water dragon	77–93 °F	50–80%
Uromastyx	85–110 °F	20–30%
SNAKES		
Ball python	77–86 °F	50–70%
Blood python	85–86 °F	50–70%
Boa constrictor	82–93 °F	70–90%
Red-tailed, Columbian, Burmese python	77–86 °F	70–90%
Corn snake	78–88 °F	20–50%
Garter snake	68–95 °F	20–30%
Green tree python	78–85 °F	80–100%
Hog Island boa	78–90 °F	80–100%
Hog-nose	80–90 °F	20–30%
King snake	73–86 °F	50–75%
Milk snake	78–90 °F	20–40%
Rainbow boa	75–90 °F	90–95%
Reticulated python	78–90 °F	80–100%
Sand Boa	77–86 °F	20–30%

(Continues)

TABLE 7-1 *(Continues)*

CHELONIANS	POTZ	HUMIDITY
Box turtle	75–84 °F	50–80%
Leopard tortoise	68–86 °F	40–50%
Painted turtle	73–82 °F	80–100%
Red-eared slider	75–90 °F	80–100%
Red-footed tortoise	80–85 °F	70–80%
Russian tortoise	78–90 °F	20–50%
Radiated tortoise (star)	80–100 °F	20–50%
Sulcata tortoise	80–95 °F	20–50%

Note: These are recommended average temperature ranges and humidity levels.

UNIT TWO

Restraint Techniques

OBJECTIVES

Upon completion of this unit, the reader should be able to:

▶ Demonstrate restraint techniques for lizards, snakes, and chelonians

KEY TERMS

autonomy

GUIDELINES FOR THE RESTRAINT OF LIZARDS

The various species of lizards have different ways of protecting themselves. Many lizards use **autonomy** (voluntary release of the tail) as a method of escape. Never grab at the tail of a lizard as a method of capture or restraint; the result will be the loss of the tail and the escape of the lizard. Tail-flicking and tail-whipping are behaviors indicating that the reptile is agitated and may become defensive or aggressive if provoked further. Iguanas are especially known for tail-whipping. Even a relatively small iguana can inflict a painful tail-whip quickly, which can cause lacerations to the face and arms of the handler. The head-bob, a quick up-and-down movement, is always a warning or a threat. Depending on the species, the dewlap (the flap of skin under the lower jaw) is erected and displayed as a threat.

All reptiles are most active when they are within their normal POTZ range, especially with regard to temperatures. It is never acceptable to induce hypothermia to have greater control of a reptile. This is inhumane and contrary to the standards of veterinary medical care.

IT IS IMPORTANT TO REMEMBER THAT...

It is important to remember that all reptiles can and will bite. Although some species are more amenable than others to handling, none should ever be considered tame or "trained." All reptile bites should be cleaned and treated immediately as a contaminated wound.

AGGRESSION IN MALE GREEN IGUANAS IS WELL DOCUMENTED

Aggression in male green iguanas is well documented. Women owners and hospital personnel have to be especially careful when male iguanas are sexually mature, as they become extremely aggressive toward women during their mensus. There have been several instances of male green iguanas attacking and inflicting severe bites and lacerations requiring emergency-room treatment, sutures, and, in some instances, reconstructive plastic surgery.

PURPOSE

- To safely handle and restrain a reptile for an examination or procedure

COMPLICATIONS

- Bites
- Scratches
- Abrasions or cuts from tail-whip

EQUIPMENT

- Assistant if the reptile is large or aggressive (All species of monitor lizards)
- Heavy towel

RESTRAINT of a Small Lizard

TECHNICAL ACTION	RATIONALE/AMPLIFICATION
1. Be prepared—lizards can move quickly.	1a. If the lizard eludes capture and restraint, do *not* make a grab for the tail or the result will be autonomy and a very unhappy client.
2. Grasp the lizard firmly around the shoulders just behind the head.	2a. If you approach from behind and hold securely, it is unlikely to be able to turn and bite.
3. Use the other hand to restrain the hind legs (**Figure 7-2**).	3a. The hind legs must be restrained to avoid being clawed or scratched.

Courtesy of Jordan Applied Technology Center, West Jordan, Utah

▲ **FIGURE 7-2** Method of restraint of a small lizard.

TECHNICAL ACTION	RATIONALE/AMPLIFICATION
4. Once the restrainer has control, the lizard is not likely to resist.	4a. This may be a survival response in small lizards, but almost all lizards will cease to struggle if they are held securely.
5. The hand that is holding the rear legs can extend the legs caudally and include the tail.	5a. Extend the limbs caudally from the pelvis, and encircle the tail with the same hand.
	5b. This provides full control of the patient for an examination or a procedure.

RESTRAINT of an Iguana

TECHNICAL ACTION	RATIONALE/AMPLIFICATION
1. Follow steps 1–5 for a small iguana. If it is larger (15 inches or longer), restraint may require an assistant to secure either the head or the hind limbs and tail.	
2. The initial capture may require a towel placed entirely over the head.	2a. Placing a towel over the head restricts vision and reaction time.

(Continues)

TECHNICAL ACTION	RATIONALE/AMPLIFICATION

3. The tail of an iguana can be tucked behind the elbow and held close to the body of the restrainer to prevent tail-whip and/or autonomy (**Figure 7-3**).

▲ **FIGURE 7-3** Restraint of the iguana holding the tail tucked under the arm and close to the body.

GUIDELINES FOR THE RESTRAINT OF SNAKES

Prior to restraining a snake, thoroughly wash your hands and forearms. All snakes hunt by scent, and if they encounter a recognizable odor, a snake is more likely to strike. Depending on the patients that were handled earlier during the day (rodents, rabbits, guinea pigs), it is also advisable to change into a clean scrub top. Movement around snakes, especially near the head, should be careful and slow, as quick movements often elicit a strike.

Some of the more aggressive snakes—for example, the emerald tree boa and the Burmese python—are likely to strike and bite the instant that visual contact is made. Before attempting restraint, it is important to understand exactly what behaviors are characteristic of the species. In all snakes, the head should be secured first. For safety, any snake, regardless of species, that is more than 5 feet long requires two restrainers—one to control the head and first third of the body, and the other to control the body and prevent the snake from wrapping around any part of the handler's body or examination table (**Figure 7-4**). Smaller snakes may be restrained by one person holding the head with one hand and loosely supporting the body with the other (**Figure 7-5**).

◀ **FIGURE 7-4** Restraint of a large snake requires more than one handler.

◄ **FIGURE 7-5** A small snake can be safely restrained by one person controlling the head.

ALL SNAKE TEETH, NOT JUST THE FANGS, CURVE CAUDALLY

All snake teeth, not just the fangs, curve caudally. If bitten, try not to react by attempting to pull away or struggle. Pulling away will tear the skin and drive the teeth deeper. Struggling causes a snake to bite harder and attempt to constrict. As difficult as it may be, remain calm and motionless. Insert a round, smooth object such as a pen or pencil between the jaws in the back of the mouth and slowly roll it rostrally (toward the nares). This will lift out the teeth and also act as a mouth gag. Scrub the bite wound thoroughly, and treat as a contaminated puncture wound.

PROCEDURE for Restraint of the Snake

TECHNICAL ACTION	RATIONALE/AMPLIFICATION
1. Quietly observe the snake for a few moments.	1a. This is done to assess the snake's attitude. You should make no move, just observe.
2. Slowly approach the snake from behind the head.	2a. Rapid movement at this point, depending on the species, may elicit a strike or an attempt to escape.
	2b. For safety, the person attempting capture for restraint should always approach from behind the head.
	2c. Movement in front of the snake will likely cause it to move away or prepare for a strike.
3. Grasp the snake just behind the head.	3a. This hold prevents the snake from biting.
4. Lift up and support the body with the other hand.	4a. The body should always be supported. If the snake is small, allow the body to drape across the other arm.
5. If the snake is large, an assistant should restrain and support the caudal two-thirds of the body.	5a. Snakes have powerful muscles designed not just for forward movement but also for constriction. If a snake is permitted to coil around a restrainer, it can be difficult to release the grip that a snake has on a person or an examination table. A snake that is startled is likely to constrict tighter. Never place a snake around someone's neck.

TECHNICAL ACTION	RATIONALE/AMPLIFICATION
11. Determine the **photoperiod**, the number of hours lights on and off.	11a. The lights should cycle for 13–15 hours during the summer months, and 9–12 hours during the winter months, to mimic the natural photoperiod of the species. Many owners rely on timer switches to turn the lights on and off.
12. Ask about the average temperature during the daytime and during the night.	12a. 13a. These temperatures should be based on POTZ guidelines and adjusted for daytime temperatures and nighttime temperatures. Owners often rely only on the lights to provide adequate heat and cool-down periods and but have no real way of measuring temperature and humidity gradients. Of the assortment of gauges available, the client should receive a recommendation to obtain one that is suitable for the habit and to monitor it closely.
13. Find out how the temperature and humidity are measured.	
14. Ask if the reptile is allowed outside of the habitat and, if so, whether it is supervised or unsupervised.	14a. Reptiles that are released to run loose in the house, or are not caged at all, especially iguanas, not only drag bacteria along with them, but they are likely to swallow foreign objects. Toy soldiers, gold jewelry and chains, and countless other items have all been surgically removed from iguanas.
15. Inquire if the reptile is taken outdoors.	15a. Reptiles often are taken outside to "enjoy the weather," especially if it is sunny. There is the benefit of natural sunlight outdoors, but also the danger of other animals, dogs especially, attacking and harassing them. In addition, the reptile may graze on toxic plants.
16. Ask whether the reptile has access to direct sunlight. Or is it filtered by window glass?	16a. Instead of taking the reptile outdoors, owners sometimes move the habitat in front of a window so it can "get some sun." Heat not only magnifies through window glass, but the glass filters out UV light. Thus, sunlight has no benefit, and the habitat can quickly become too hot and kill the reptile.
17. Find out if the patient has access to live plants. If so, what kind of plants?	17a. 18a. Many household plants are toxic to pets, including reptiles.
18. Ask the owner if he/she has ever seen the reptile consuming any part of the plants.	
19. Inquire about how the water is provided and if there is a place for the reptile to soak.	19a. 20a. 21a. Water availability is essential to all reptiles. Some smaller species of lizards may lap drops of water misted into the cage or drink from a bowl; others, such as chameleons, need a small waterfall, as they do not drink from standing water. Some reptiles, the iguana for example, normally defecate in water. All snakes should have a bowl large enough so they can soak, but reluctance to leave the water bowl may be a sign of illness. Water quality must be monitored for aquatic turtles. All water should be dechlorinated and filtered.
20. Ask if the water is dechlorinated or if anything is added to the water.	
21. Ask if the reptile spends a lot of time soaking.	

(Continues)

TECHNICAL ACTION	RATIONALE/AMPLIFICATION
22. Determine when the reptile last shed and if there was difficulty with the shedding process.	**22a.** Difficulty shedding can indicate a lack of humidity or another health concern.
23. Ask if the owner has noticed any changes in the reptile's behavior.	**23a.** As with any animal, changes in behavior can indicate many things. It could indicate stress, sexual maturity, or an illness. Gather as much information as possible regarding any change that the owner is reporting.
24. Inquire about how often the habitat is cleaned and what cleaning agents are used.	**24a.** Some cleaning agents can be toxic when used in reptile habitats.
25. Ask if the owner has administered any over-the-counter medicines or treatments advised by the pet store. If so, what was it, how it was given, how much, and how frequently?	**25a.** Clients often return to the pet store seeking medical advice. Many times they are sold products that have little or no benefit, and they delay qualified veterinary attention in the hope that an over-the-counter (OTC) product will help or cure the reptiles. When the patient is presented, it may be critical. Many OTC products and home remedies interfere with necessary and prescribed medications.
26. Ask if the owner is aware of the risk of Salmonella, what it is, and how it is transmitted to people.	**26a.** All clients should be advised of the risks of Salmonella, especially for young, elderly, and immune-compromised members in the household. Many practices ask clients to sign handouts indicating that they have been given a copy of this information and have had their questions answered. This documentation should become a part of the permanent medical record.

Physical Examination

OBJECTIVES

Upon completion of this unit, the reader should be able to:

▶ Perform a basic examination of lizards, snakes, and chelonians

KEY TERMS

carapace

episodic

gaping

rostrum

© Kuttlevaserova Stuchelova/Shutterstock.com

GUIDELINES FOR BASIC PHYSICAL EXAMINATION OF REPTILES

A complete and thorough patient history provides the basis for a reptile physical examination. Standard physical exam forms may be used with some modifications, and some areas, as with birds, are excluded from the exam. For instance, the temperature isn't taken, and the respiration rate may be difficult to evaluate because reptiles often are **espisodic** breathers in that they may demonstrate a regular breathing pattern and then not breathe at all for a short time. The heart rate may be difficult to detect because of the scales or plating of the chelonian shell, and there are no palpable lymph nodes.

PURPOSE

- To evaluate the patient for medical concerns and with the information from the patient history, the veterinarian will be able to determine a treatment or wellness plan.

COMPLICATIONS

- Restraint technique
- Aggressive or difficult patient

EQUIPMENT

- Assistant for restraint, if required
- Towel
- Gram scale with containment bucket
- Human infant cradle-type scale (pound/ounce weight has to be converted to gram weight)
- Pillow case or "snake bag"
- Stethoscope
- Damp towel
- Tongue depressor sticks

PROCEDURE for a Reptile Examination

TECHNICAL ACTION	RATIONALE/AMPLIFICATION
1. Assess overall appearance and body condition.	1a. In malnourished lizards, the pelvic bones become prominent and the tail appears shriveled and thin. Obese lizards often have a distended abdomen with palpable fat pads along the body. The tail base also is enlarged with fat pads. Snakes that are undernourished will exhibit decreased muscle mass with prominent dorsal processes and ribs. Obese snakes have fat pads along the lower third of the body. It is difficult to assess body condition in chelonians, but generally, when lifted, an undernourished turtle or tortoise will feel "empty." There may be little or no fleshy parts visible. Obese chelonians sometimes have difficulty retracting their legs entirely because of fat.

(Continues)

TECHNICAL ACTION	RATIONALE/AMPLIFICATION
2. Record observations on respiration rate.	**2a.** Prior to handling, the following should be observed: Does the patient take regular breaths? Is the breathing episodic or fairly regular?
	2b. Note any audible breath sounds or **gaping** (open-mouthed breathing), which may indicate an upper respiratory infection. Snakes that gape may also exhibit excessive salivation. Exhaling produces a sound not unlike a hiss.
	2c. Reptiles do not have a diaphragm to assist with respiration. Abdominal and intercostal muscles move the air in and out of the lungs. (In snakes, only the right lung is inflated.)
3. With restraint, look closely at the eyes and aural openings.	**3a.** The eyes of a healthy reptile should be clear and open. The eyes of sick reptiles typically are partially closed, and there may be an ocular discharge. **Figure 7-11** demonstrates a lizard that is in an extremely poor condition.
	3b. The aural openings should be checked for swelling, disruption of scales around the opening, or any discharge. Swelling could indicate an abscess or inflammation. **Figure 7-12** is an example of a large aural abscess in a red-eared slider turtle. This is also a prime spot for mites to infest. Mites burrow under scales, disrupting the pattern and, with closer examination, can also be found within the aural cavity. Mites feed on blood and can drastically debilitate a reptile.

Courtesy of Bev and Dan Ring

▲ **FIGURE 7-11** This leopard gecko is in very poor health. With dedicated care and correct management of the habitat and diet, the gecko slowly returned to health.

Courtesy of Eric Klaphake, DVM

▲ **FIGURE 7-12** A large aural abscess on a red-eared slider turtle.

4. Examine the **rostrum** (tip of the nostrils and upper mouth) or beak if a chelonian.	**4a.** Damage to the rostrum often occurs when a reptile is housed incorrectly; the habitat may be too small for the activity level, and the patient frequently charges into the walls, either in an attempt to escape (fright) or aggression toward another reptile housed in close proximity.

(Continues)

TECHNICAL ACTION	RATIONALE/AMPLIFICATION

▲ **FIGURE 7-13** Rostral abrasion and trauma on Basilisk lizard.

5. Examine the mouth and oral cavity.

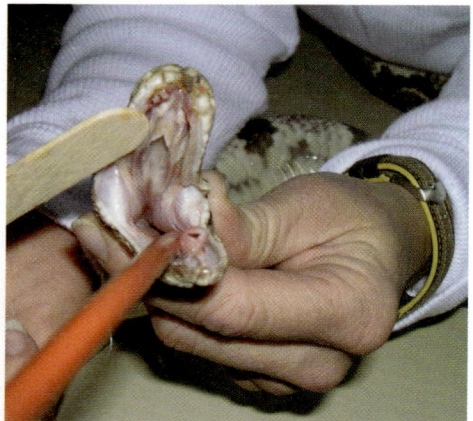

▲ **FIGURE 7-14** Examining the mucous membranes of a snake for areas of discoloration and abnormal swelling.

6. Attempt to auscultate heart and respiratory sounds.

▲ **FIGURE 7-15** The use of a damp towel wrapped around the thorax amplifies sound and reduces interference from scale noise and can help hear heart sounds.

Figure 7-13 shows a male basilisk lizard with bulbous rostral damage. Basilisks stand and run on their hind legs to escape. The basilisk is frequently frightened, sometimes deliberately because it is "fun" to watch it stand and run. The housing and (and human behavior) is inappropriate to accommodate this species.

Snakes rarely present with rostral damage, but when it occurs, it usually is the result of an overly aggressive strike in a feeding enclosure that is too small. The beak of a chelonian should be examined for splits, cracks, hardness, and obvious fractures. The bite also should be assessed. Chelonians may present with overgrown or malformed beaks.

5a. The oral cavity should be examined, looking for any signs of discoloration or abnormal swelling. The normal color of mucous membranes in reptiles is pale pink to white and does not suggest anemia as it would in a mammal (**Figure 7-14**). The tongue of healthy snakes and monitor lizards normally flicks in and out.

6a. Heart rates in reptiles are not as easy to determine as in other species. The rates vary by species, and external temperatures also affect the heart rate and the patient's general metabolic activity. Reptile hearts are three-chambered and are located more dorsally than those of mammals and auscultation is difficult because of the scales and/or solid breast plates. Wrapping a damp towel around the thorax of a lizard amplifies the sounds and decreases interference from scale noise (**Figure 7-15**).

(Continues)

TECHNICAL ACTION	**RATIONALE/AMPLIFICATION**

7. Evaluate the general appearance of the integument.

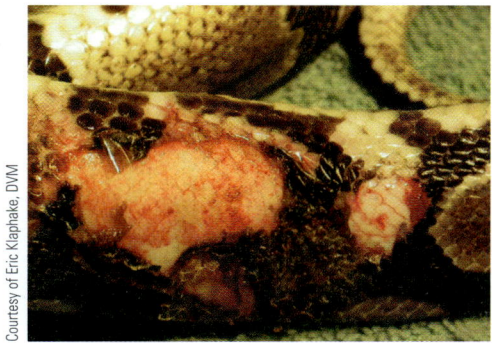

Courtesy of Eric Klaphake, DVM

▲ **FIGURE 7-16** Bites from a live rat that was left in an enclosure with a snake.

7a. Examine the skin and scales for evidence of mites and dysecdysis. Lizards and snakes frequently present with thermal burns from faulty heat rocks and the close proximity of basking lights. Reptiles do not have a withdrawal reflex and cannot perceive that a light or a rock is too hot, so they continue to lie on top of a hot item, often burning through the skin and underlying muscle tissue. Snakes that are offered live prey and are left unsupervised are often chewed on by the rat or mouse that is left in the enclosure (**Figure 7-16**).

Examine the condition of the shell of a turtle or tortoise, and look for flaking areas of the **carapace** (the top half of the shell) and redness on the underside of the plastron (lower half of the shell). Evaluate the visible skin where the legs join the body, for signs of edema or dehydration. Chelonians shed their skin in small pieces from the limbs, neck, and tail over a short period of time.

8. Examine each digit, nails, and tail.

8a. Note any missing toes or nails. When pieces of a retained shed dry, the dead skin tightens and cuts off the blood flow to the toe, or possibly the end of the tail. Reptiles may present with a broken tail, or a tail that has been lost as a result of trauma or autonomy. Generally, missing tails will regenerate over several months. The re-grown tail has a distinctly different appearance and color.

9. Examine the vent/cloaca.

9a. The vent (cloaca) should be clean and dry with no evidence of mucus, inflammation, or abnormal swelling. Some species of snakes have small spurs on either side of the vent. These are small bones covered in keratin and are used in courtship and mating. (The relative size of the spurs is not a reliable method of sex determination.

10. Obtain an accurate weight in grams.

10a. Underweight reptiles often feel "empty" when they are picked up. This is especially true with chelonians. Small reptiles may be weighed in a lidded container on a gram scale. If the reptile is too large for the container, some clinics use a human infant scale with a cradle. The weight given is usually in pounds and ounces, and it must be converted to grams. (Divide the pound weight by 2.2 to obtain the weight in kilograms.) If it is a large tortoise, it can be weighed by placing it on a small-animal scale. Snakes usually are weighed in a "snake bag" or pillowcase.

UNIT SIX

Care of the Hospitalized Patient

OBJECTIVES

Upon completion of this unit, the reader should be able to:

▶ Provide the correct environment for the hospitalized patient

GUIDELINES FOR CARE OF THE HOSPITALIZED PATIENT

Veterinary technicians must have a good understanding of the requirements of individual species and be able to provide a hospital environment that is supportive and therapeutic. Many critically ill reptiles are housed in incubators where temperature and humidity are easily controlled. These self-contained units are easily cleaned and disinfected and provide a good escape-proof environment (**Figure 7-17**).

Courtesy of Jordan Applied Technology Center, West Jordan, Utah

▲ **FIGURE 7-17** Climate controlled incubators make ideal housing for hospitalized reptiles.

Other patients may be suitably housed in simple glass aquaria with screened lids. In any case, the appropriate lights, heating, and photoperiods have to be provided. Accommodation must also be provided for aquatic species: water quality, aeration, filtration, and water temperature.

Many patients require supportive feeding, fluid therapy, and medications. In addition, there may be radiographs, anesthesia, and surgical procedures. Reptile medicine is a growing part of total patient care in a small-animal practice. Acquiring new skills for these unique patients can be challenging but highly rewarding.

UNIT SEVEN

Blood Collection

OBJECTIVES

Upon completion of this unit, the reader should be able to:

▶ Perform venipuncture and blood collection

KEY TERMS

hemipenes

GUIDELINES FOR BLOOD COLLECTION IN REPTILES

Prior to any blood draw, determine the tests to be performed and if they will be done in-house or sent to an outside exotics laboratory for testing. If samples are to be sent out, always call the lab first and ask about specific requirements for the desired tests, type of blood container, and sample volume. Blood volume for most species is between 5% and 8% of body weight. Approximately 10% of this then can be collected at one time. Many outside labs request little as 0.5 ml of blood.

If a patient is already dehydrated or critically ill, the volume of blood withdrawn is often replaced with the same amount of electrolytes or 0.9% saline to prevent volume depletion. This is especially true for very small reptiles.

Venipuncture sites in lizards include the ventral coccygeal vein and the ventral abdominal vein. The most commonly used blood collection sites in snakes are the ventral coccygeal vein and cardiac puncture. Venipuncture sites in chelonians include the jugular vein, the dorsal and ventral coccygeal veins, and the subcarapacial vein.

PURPOSE

- To obtain an adequate sample for blood and biochemistry tests with the least amount of trauma and stress to the patient
- To collect samples for either in-house testing or preparation for an outside diagnostic laboratory.

COMPLICATIONS

- Inadequate restraint
- Not preparing a clean site for venipunture
- Introduced bacteria from contaminated puncture site
- Trauma at puncture site

EQUIPMENT

- Dilute chlorhexidine (Novalsan®) solution
- Warmed water
- 25–27-gauge needle
- 1cc syringe
- Lithium heparin (green top) tube
- Warmed electrolyte solution or 0.9% saline if volume replacement is required
- Assistant for restraint

CHELONIANS WILL TOLERATE BEING PLACED IN DORSAL RECUMBANCY

Chelonians will tolerate being placed in dorsal recumbency for short periods, provided that they are turned over slowly and gently. Alternatively, you should follow the steps below and access the dorsal tail vein to avoid turning the turtle onto its back. Some of the larger tortoises may have to be anesthetized prior to the procedure simply because their strength makes restraint and exposing the tail difficult. If they are anesthetized, someone must monitor the patient and recovery. All reptiles possess a strong righting reflex and will make every attempt to right themselves unless they are adequately restrained.

PROCEDURE for a Coccygeal Venipuncture in Lizards, Snakes, and Chelonians

TECHNICAL ACTION	RATIONALE/AMPLIFICATION
1. Have all equipment ready.	
2. Restrain the patient in dorsal recumbency.	2a. Read the sidebar notes on chelonians.
3. Thoroughly clean the puncture site with dilute chlorhexidine and a warm water rinse.	3a. Reptiles often drag their tails through fecal material and contaminated water.
4. Insert the needle between the scales at the midline of the tail approximately one-third of the distance from the vent (**Figure 7-18**).	4a. This is to avoid puncturing the **hemipenes**, the paired sacs in lizards and snakes that empty into the cloaca and are brought together during copulation.

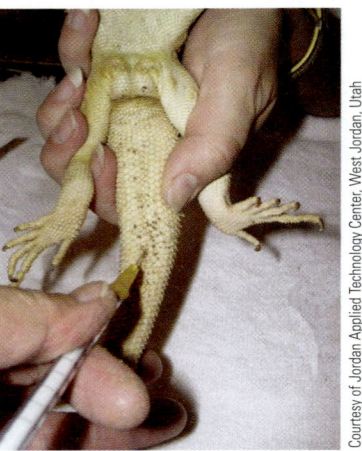

Courtesy of Jordan Applied Technology Center, West Jordan, Utah

▲ **FIGURE 7-18** Drawing blood from the coccygeal vein a small lizard.

5. Hold the needle perpendicular to the tail and insert to the vertebral bone.	5a. The needle should not enter bone; contact with bone is a marker.
6. Draw back slowly on the syringe plunger and back the needle out until blood flashes in the hub.	6a. This method backs the needle directly into the vein. Be careful not to withdraw too far or the needle will be pulled entirely though the vein.
7. Collect the sample and withdraw the needle.	
8. Remove the needle from the syringe and transfer the sample directly from the syringe to the green top (lithium heparin) tube.	8a. Because of the narrow gauge of the needle, forcing the sample back through it and into the tube is likely to lyse the red blood cells. Red blood cells in reptiles, like birds, are large, oval and nucleated.
	8b. Samples in lithium heparin tubes do not have to clot prior to separating the sample, and are preferred for biochemistry and electrolyte analysis. If sending the sample out to an exotics lab, always check with them for their sample requirements *prior* to a blood draw.
9. Apply pressure to the puncture site for at least 60 seconds or until bleeding stops.	

PROCEDURE for Ventral Abdominal Venipuncture in Lizards and Snakes

TECHNICAL ACTION	RATIONALE/AMPLIFICATION
1. Follow steps 1–3 above.	
2. With the index finger, apply pressure on the cranial aspect of the vein.	
3. Hold the needle and the syringe almost parallel to the puncture site, and insert the needle between the scales exactly at the midline.	3a. Holding the syringe at a greater angle will either penetrate the vein or miss it entirely.
4. This is a blind stick; gently move the needle until the vein is accessed. Do not withdraw the needle.	
5. When blood flashes in the hub, follow steps 6–9 above.	

PROCEDURE for Cardiac Puncture in a Snake

TECHNICAL ACTION	RATIONALE/AMPLIFICATION
1. Clean venipuncture site per steps 1–3.	
2. Restrain the patient dorsally, or have two people hold it aloft.	2a. If the patient is held aloft, it often is easier to visualize the heartbeat, as it tends to "drop" closer to the ventral surface. If restrained dorsally, the heart will be deeper in the body cavity.
3. Use careful, patient observation to visualize the heartbeat.	3a. The heart rate of a snake is substantially lower than that of mammals, and several seconds of observation may be required to locate the heart.
	3b. The heart can be seen or felt in the distal end of the first third of the length of the snake's body.
4. Once visualized, mark the site with a non-toxic felt-tip pen.	4a. The marks should be made lightly, only enough to determine the area of the heart. A dark color is best because the ventral surface of almost all snakes is a pale cream color to white. A non-toxic marker pen is necessary because reptiles absorb toxins through their skin.
5. Isolate the site between the thumb and the index finger.	5a. Needle movement potentially could lacerate the heart. It could also cause damage by completely penetrating the walls of the heart and lead to death of the patient.
6. Insert the needle straight into the heart, avoiding needle movement as much as possible.	
7. Follow steps 5–9 above, and maintain pressure at the puncture site until all bleeding has stopped.	

CARDIAC PUNCTURE

A cardiac puncture can be performed in a soft-shelled turtle. The turtle should be restrained dorsally and the puncture site cleaned. The heart is located slightly left of midline and approximately one-quarter caudally of the length of the plastron. The needle should be inserted perpendicular to the plastron.

PROCEDURE for Subcarapacial Venipuncture in a Large Chelonian

TECHNICAL ACTION	RATIONALE/AMPLIFICATION
1. Follow steps 1–3 above.	
2. Restrain tortoise from walking.	2a. Restrain either by holding a foreleg with the opposite hand or have an assistant restrain by holding the tortoise with both hands around the shell.
3. When the head is retracted into the shell, the approach of the needle should be at a 60° angle just caudal to the skin attachment to the carapace (**Figure 7-19**).	3a. When the head is retracted, this vein is easier to access. This is a good method for a large tortoise.

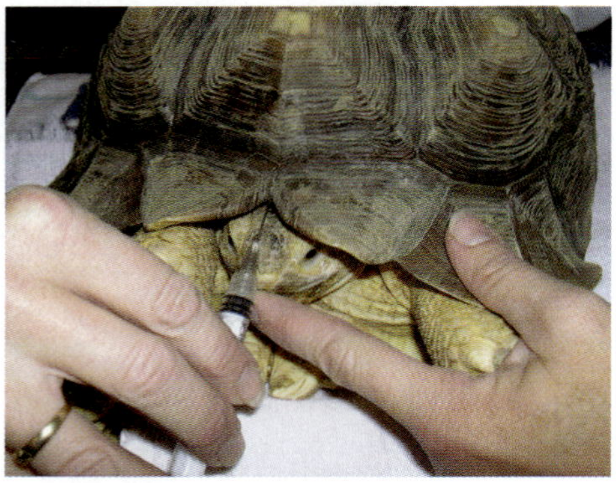

Courtesy of Jordan Applied Technology Center, West Jordan, Utah

▲ **FIGURE 7-19** Approach to the subcarapacial vein in a tortoise. The head should be retracted into the shell.

| 4. Follow steps 6–9 above. | |

PROCEDURE for a Jugular Venipuncture in a Chelonion

TECHNICAL ACTION	RATIONALE/AMPLIFICATION
1. The restrainer should have a firm grip on the neck to prevent the head from retracting into the shell. (Anesthesia may be required for large chelonions.)	1a. The head can retract quickly into the shell if not firmly restrained. If the needle has already been placed severe laceration of the vein may occur or the needle may break off in the vein.

(Continues)

TECHNICAL ACTION	RATIONALE/AMPLIFICATION

1b. Holding the head from top to bottom (dorsal/ventral) also prevents the turtle from being able to bite, and it keeps the hands clear of the venipuncture site.

2. Follow steps 1–3 above.

3. Grasp the head of the patient firmly with the index finger and thumb of the opposite hand.

4. Extend the head and turn it slightly away from you, exposing the jugular vein (**Figure 7-20**).

4a. The right jugular vein is more prominent. Jugular veins are located dorsally, and turning the head slightly to the left will give better exposure.

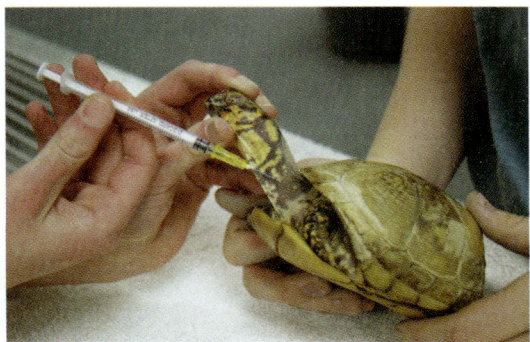

▲ **FIGURE 7-20** Restraint and positioning for drawing blood from the jugular vein of a turtle.

5. Follow steps 6–9 above. The person restraining the extended neck should not release it until all bleeding has stopped.

5a. If restraint of the neck is released too soon, it will be very difficult to re-extend the neck after the procedure because the patient will have retreated firmly into the safety of the shell.

UNIT EIGHT

Medication Administration

OBJECTIVES

Upon completion of this unit, the reader should be able to:

▶ Perform the correct administration of medication to the reptile

KEY TERMS

renal portal system

GUIDELINES FOR ADMINISTRATION OF MEDICATION

Renal Portal System of a Lizard

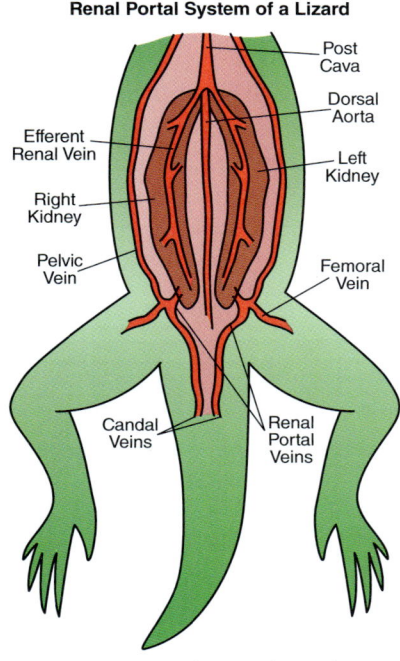

▲ **FIGURE 7-21** The renal portal system in reptiles.

Medications can be administered orally (gavage method) or by injection. Generally, injectable medications are used more frequently than oral medications because of the difficulty in administering oral medication, the unpredictable absorption rate, and the metabolism of the drug. Injections can be administered SQ (subcutaneously), but this can be painful to the patient as there is little space between the skin and musculature. More frequently, injections are given IM (intramuscularly).

All IM injections in reptiles should be given in the forelegs or, in snakes, the dorsal muscles either side of the vertebral column in the first third of the length of the snake's body. If injections are given in the rear legs or caudal aspect of the body, the **renal portal system** moves blood from the caudal third of the body into the kidneys first and then into general circulation, effectively filtering out medication before it can be absorbed (**Figure 7-21**).

PURPOSE

- To deliver a therapeutic dose of medication

COMPLICATIONS

- Delivery into a rear limb and the medication being filtered out through the renal portal system
- Difficulty in assessing efficacy of dose with oral medications because of metabolic variables

EQUIPMENT

- 22–25-gauge needle
- Appropriate-sized syringe
- Metal gavage tube for oral medication or supplemental feeding

PROCEDURE for Administering IM Injections

TECHNICAL ACTION	RATIONALE/AMPLIFICATION
1. Follow all normal protocol for administering injections safely.	1a. In choosing the injection site, it is important to keep the renal portal system in mind and inject only into the muscles of the forelimbs or cranial first third of the snake's body.
2. Clean injection site.	
3. Use a 22–26-gauge needle, and always insert between the scales.	

PROCEDURE for Administering Oral Medications

TECHNICAL ACTION	RATIONALE/AMPLIFICATION
1. If the medication is in tablet or capsule form, first make it into a solution.	1a. Separate the capsule or grind the tablet with a mortar and pestle and make a solution by adding water.

(Continues)

TECHNICAL ACTION	RATIONALE/AMPLIFICATION
2. Draw the medication into the syringe and attach it to metal gavage tube.	
3. Insert the ball tip in the side of the patient's mouth.	**3a.** Use the ball tip of the gavage tube to pry open the mouth from the side, lifting up the maxilla rather than pushing down on the mandible. This is more likely to encourage the reptile to open its mouth. The glottis is clearly visible in reptiles. Pass the tube behind it to avoid aspiration.
4. Depress the plunger and deliver the medication.	
5. Slowly remove the gavage tube.	

A RED RUBBER CATHETER IS USED

A red rubber catheter is used sometimes, but the reptile can bite through it. The portion within the esophagus will have to be removed with forceps, potentially causing trauma. To avoid this when using a rubber catheter, place a roll of white zonas tape vertically in the mouth and pass the tube through the center of the roll. There is a danger of being bitten when the roll of tape is removed.

GUIDELINES FOR FLUID THERAPY

Many reptile patients are presented with dehydration because of a lack of humidity and no access to a water source that is large enough for them to climb into and soak. This, in turn, can lead to blockage of the intestinal tract, fecal impaction, and loss of appetite. If the dehydration is not too severe, the simplest way to rehydrate a reptile is to soak the patient in a warm water bath for 10–15 minutes three or four times a day. Unless additional supportive treatments require hospitalization, owners can do this at home. They should be advised, because of the risk of Salmonella, to obtain and use a dedicated tub for soaking. The reptile should not be placed in a household sink or a bathtub, as soaking often stimulates defecation. (Many reptiles normally defecate only in water.)

The bath only has to be plain warm water from the tap, and the container must be large enough for the reptile to submerge its entire body but shallow enough for its head to remain out of the water. The client should also be advised to supervise the reptile during the soaks to prevent it from climbing out or drowning. It is not advisable to put the reptile in a plastic or rubber tub with a clamped-on lid. Water in many species is also absorbed through the cloaca.

CLINICAL FLUID THERAPY IS NECESSARY

In severe cases, clinical fluid therapy is necessary. Reptiles can be given fluids intravenously, by subcutaneous injection, or by bolus into the celomic cavity. Gravely ill patients may require placement of an IO (intraosseous) catheter. This procedure requires general anesthesia or, at the very least, a local anesthetic, and it should not be attempted without consultation and or direct supervision of a person who is well qualified and experienced in reptile medicine.

PURPOSE
- To rehydrate or maintain hydration status
- To correct electrolyte imbalances
- To encourage defecation

COMPLICATIONS
- Potential rupture of internal organs with ICe administration

EQUIPMENT
- Dedicated tubs of various sizes for soaking
- Assistant for restraint if giving fluids SQ or ICe
- 22–25-gauge needle with appropriate-sized syringe for the volume of fluids to be delivered

ADMINISTRATION of Fluid Therapy

TECHNICAL ACTION	RATIONALE/AMPLIFICATION
1. Place patient in a tub of warm water for approximately 15 minutes three or four times a day.	
2. Flush out the tub and disinfect, then allow to air-dry.	2a. Maintenance fluids are calculated at 5–10 ml/kg/day. Therapeutic/restorative total volume will be greater and may be 15–25 ml/kg/day. The total daily volume should be divided into three or four treatments per day.
3. If delivering fluids by injection, obtain the patient's weight.	
4. Calculate the correct amount of fluids for the patient.	
5. Clean the delivery site.	5a. In lizards, the fluids can be given between the lateral scales or in front of the hind leg and directed toward the opposite shoulder.
	5b. Chelonians receive SQ fluids in the inguinal fold or the ventral neck flap.
	5c. Fluids can be given to a snake in the last fourth of the coelmic cavity.
	5d. All needle punctures should be made between the scales.
6. Administer fluids as a bolus.	

UNIT NINE

Anesthesia

OBJECTIVES

Upon completion of this unit, the reader should be able to

▶ Safely perform reptile anesthesia

GUIDELINES FOR ANESTHESIA IN REPTILES

The considerations for reptile anesthesia are different from those for other species. Primarily, it must be remembered that they are ectotherms; therefore, they will be slower to induce, more difficult to maintain, and have a prolonged recovery time. Reptiles do not have a diaphragm, and respiration is dependent on abdominal and intercostal muscles to move air in and out of the lungs.

In addition, reptiles are episodic breathers; they may take a couple of breaths and then stop breathing for a short time. Reptiles commonly become apneic when they are completely anesthetized. The patient should be ventilated manually several times in sequence to inflate the lungs similar to normal respiration and help maintain the anesthesia level. Normal respiration for an anesthetized reptile is two to four breaths per minute.

The reptile heart has three chambers; two atria and one ventricle. The ventricle has three subchambers with shunts that force the blood to the body and lungs (**Figure 7-22**). These valve-like shunts can function independently or in unison. The cardiac mechanism has a direct effect on oxygen saturation levels and the elimination of anesthetic gas during surgery and recovery. A reptiles can be induced by placing it in a sealed plastic bag and inserting the breathing tube into the bag and taping it around the seal. An anesthesia induction chamber may also be used prior to intubation. Induction will be slow, usually around 20 minutes. At no time should the patient be left unsupervised.

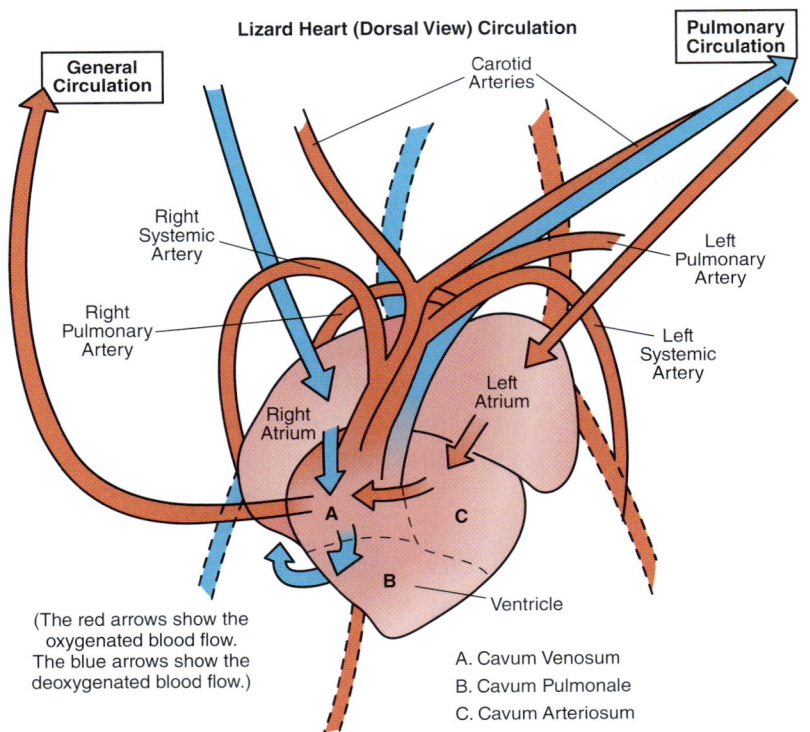

▲ **FIGURE 7-22** Structures of the reptile heart.

Injectable anesthetics can be used; however, they lower cardiac output, causing slower blood distribution, which results in a slower absorption rate. When using injectable anesthetics in reptiles, an increased time for drug effectiveness is expected. Administering a second volume will not shorten induction time but, rather, will lead to anesthetic overdose and death. Most injectable anesthetics are administered intravenously. This improves induction and recovery time and is less irritating to the patient. If given ICe, there is a risk of lacerating internal organs. In addition, some anesthetic agents can be highly irritating and cause tissue damage.

With an inhalant anesthesia, reflexes are lost in this order: mid-body (and righting reflex), then cranially, then caudally, with the tail the last to lose all reflexes. Recovery occurs in the opposite direction. (The tail should be protected and restrained during recovery.)

Intubation of reptiles is easily achieved as the glottis is clearly visible; in lizards and chelonions it is behind the tongue, and in snakes it is rostral and appears as a "tube." Uncuffed endotracheal (ET) tubes should be used because reptiles have incomplete tracheal rings. Inflation of a cuffed ET tube can cause trauma to the trachea. Lizards are straightforward to intubate, but you have to be aware of differences. In snakes, only the right lung inflates. The chelonion patient poses a different problem: Chelonians have short bifurcated tracheas. ET tubes should be placed at the top of the bifurcation to avoid inflating only one lung. When chelonians breathe, the abdominal and neck muscles assist with air flow. When a turtle or a tortoise retracts its head, it stops muscle movement and halts respiration. An anesthetized chelonion should have its head and neck fully extended to prevent respiratory compromise.

Isoflurane induction and maintenance rates are 1.5%–3.0%. Sevoflurane is more variable, and induction rates are higher. Maintenance is more difficult to achieve and the anesthetist will have to monitor anesthetic depth and adjust the flow of gas accordingly.

The oxygen flow rate for both agents is 50–100 ml/kg.

PURPOSE

- To deliver anesthesia to a patient for a surgical procedure, radiographs, or restraint.

COMPLICATIONS

- Slow induction and recovery
- Inadequate ventilation during the period of anesthesia
- Unreliable monitoring perimeters
- Mucous plugs in ET tube

EQUIPMENT

- Clear plastic bag or induction chamber
- Tape for securing the breathing tube and bag
- Appropriately sized uncuffed ET tubes
- Stethoscope, esophageal stethoscope, or Doppler
- Pre-warmed incubator for recovery

PROCEDURE for Anesthetizing a Reptile

TECHNICAL ACTION	RATIONALE/AMPLIFICATION
1. Induce patient with inhalant anesthesia as described above.	
2. Intubate and transfer the patient to a non-re-breathing system attached to the anesthesia machine.	
3. Place the esophageal stethoscope or Doppler.	3a. When placing the esophageal stethoscope, care must be taken not to enter the stomach. The location should be adjusted so it is situated directly over the heart, where sounds are most audible.
	3b. The Doppler probe may be placed over the heart, carotid artery, or coccygeal artery.
	3c. Pulse oximeters are not the most reliable in assessing vital signs in reptiles. The primitive brain stem of reptiles continues to function, producing recognizable sounds of cardiac output when, in actuality, the patient may be unrecoverable.
4. Transfer the patient to the surgical table, and position it for the procedure.	
5. Ventilate the patient several times in sequence to mimic normal respiration.	5a. If the patient is resistant to manual ventilation, the ET tube should be checked for mucous plugs, and replaced if obstructed.
6. Check for tail movement frequently.	6a. During recovery or when the plane of anesthesia becomes light, the tail is the first to regain muscle control.
7. Recovery: Turn off the inhalant agent, ventilate two to four times with pure oxygen, repeat after 2 minutes, extubate, and place the patient in a pre-warmed incubator.	7a. Pre-warming the incubator assists in recovery time.
8. Monitor recovery status using the tail reflex and righting instinct.	8a. The tail should be observed for movement. There may be none, but turning the patient onto its back may trigger the righting reflex. (Do not leave the patient on its back.) To determine if the patient is unrecoverable, open the mouth and observe the glottis to see if it is open or closed. The glottis of a deceased reptile remains open. Rescue may be attempted by re-intubating and manually ventilating.

Surgical Preparation

OBJECTIVES

Upon completion of this unit, the reader should be able to

▶ Prepare the patient for surgical procedures

GUIDELINES FOR SURGICAL PREPARATION OF THE REPTILE PATIENT

Reptiles are prepared for surgery using the same aseptic techniques as for any other species. Chlorhexadine scrub is preferred because it is less irritating. No scales should be removed at the surgical site. Minor surgery (abscess removal) may be performed on the wet table; other procedures will occur in the OR with full surgical prep of not only the patients but also the surgical team. If the procedure is to be in the OR, the room should be set up the same as for any surgery. The only difference is that the surgeon may prefer to use a laser or electro-cautery device for cutting and wire or surgical staples for closing. Always know what the procedure will be and the specific surgical equipment needed or preferred.

Because of the prolonged induction time for reptile patients, always confirm the anticipated time that the procedure will begin and choose the preferred method of induction for the procedure well ahead of time.

UNIT ELEVEN

Radiology

OBJECTIVES

Upon completion of this unit, the reader should be able to:

▶ Correctly position the patient for various radiographic views

RADIOLOGY TECHNIQUES

Radiographs are extremely helpful in diagnosing problems. High-detail film with compatible cassettes will produce the best results. Most radiographs can be taken without anesthesia or chemical restraint, just patience. Normal views for lizards and snakes are dorsal/ventral (D/V) and lateral (**Figures 7-23** and **7-24**).

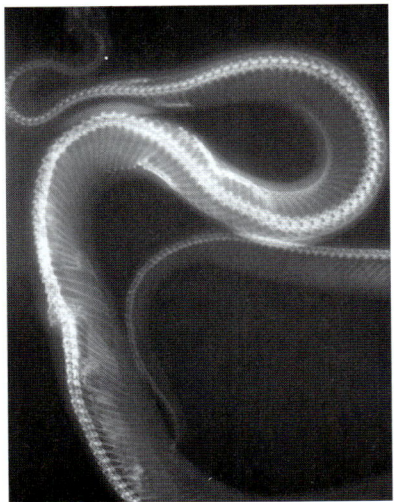

▲ **FIGURE 7-23** D/V view of a snake.

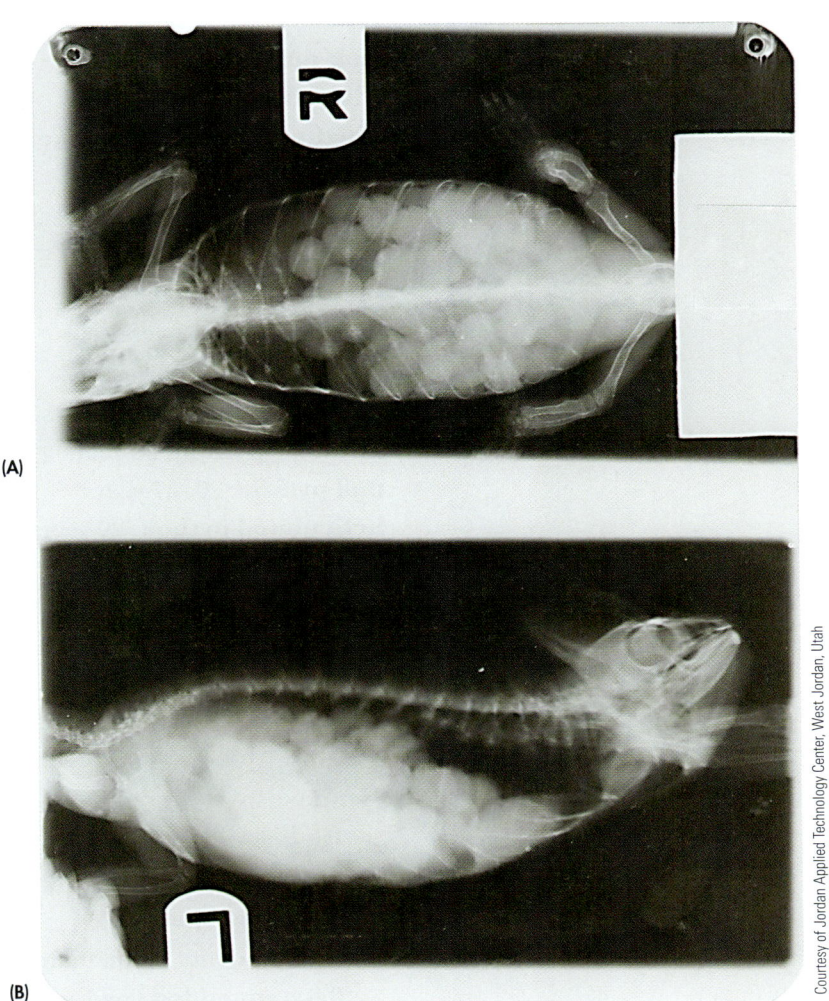

(A)

(B)

Courtesy of Jordan Applied Technology Center, West Jordan, Utah

▲ **FIGURE 7-24** (A) Lateral view of a lizard. (B) D/V view of a lizard.

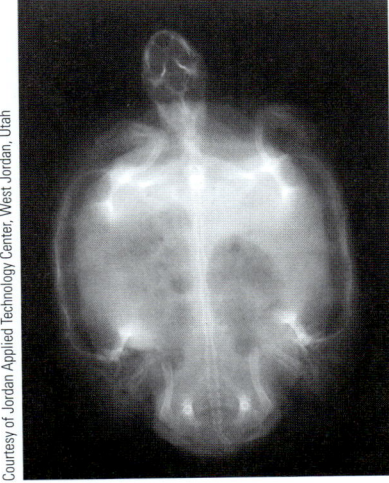

Courtesy of Jordan Applied Technology Center, West Jordan, Utah

▲ **FIGURE 7-25** D/V view of a turtle.

Snakes should be uncoiled for radiographs. Often, they can be placed directly onto the film cassette. Depending on the length of the snake, several films may have to be taken in sequence, one body length at a time. The films should be numbered sequentially, beginning with the cranial end as film number one and moving the sequence of films caudally. If it is a large snake, restraint assistance is required. Snakes *sometimes* can be persuaded to enter a length of clear plastic tube, making it easier to obtain a fairly straight position; otherwise, the snake should not have any length of its body overlapping another.

Chelonians are the easiest to position for a radiograph. Simply place the patient on top of the cassette for a D/V view. If a lateral or cranial/caudal view is needed, place the patient on top of an object with all four legs suspended (**Figure 7-25**).

FAST FACTS

Reptiles

Lifespan (Colubrids)

- **Cornsnake:** 32 years
- **Ratsnakes:** 22 years
- **Hog-nose snake:** 19 years
- **Kingsnakes:** 19–44 years
- **Northwestern garter snake:** 15 years
- **Boa constrictor:** 40 years
- **Solomon Island boa:** 16 years
- **Rubber boa:** 26 years
- **Emerald tree boa:** 31 years
- **Rainbow boa:** 31 years
- **Carpet python:** 26 years
- **Green tree python:** 19 years
- **Burmese python:** 28 years
- **Ball python:** 20–47 years
- **Reticulated python:** 29 years
- **Box turtle:** 20 years
- **Red-eared slider:** 20 years
- **Red-foot tortoise:** 20–30 years
- **Softshell:** 12–15 years

Reptile Diets

- **Pythons:**
  - **Diet:** Rodents, small mammals, chicks
  - **Recommended Feedings:** 2–3 weeks
- **Boas—larger species:**
  - **Diet:** Rodents, small mammals, chicks
  - **Recommended Feedings:** 2–3 weeks
- **Boas—smaller species:**
  - **Diet:** Rodents
  - **Recommended Feedings:** 1–2 weeks
- **Ratsnakes:**
  - **Diet:** Rodents, chicks
  - **Recommended Feedings:** 2–3 weeks
- **Cornsnakes:**
  - **Diet:** Rodents, checks
  - **Recommended Feedings:** 1–2 weeks
- **Gopher snakes:**
  - **Diet:** Rodents, chicks
  - **Recommended Feedings:** 1–2 weeks

- **Pinesnakes:**
  - **Diet:** Rodents, chicks
  - **Recommended Feedings:** 1–2 weeks
- **Kingsnakes:**
  - **Diet:** Amphibians, rodents, fish, small lizards
  - **Recommended Feedings:** 1–2 weeks
- **Watersnakes:**
  - **Diet:** Amphibians, rodents, fish, small lizards
  - **Recommended Feedings:** 1–2 weeks
- **Garter snakes:**
  - **Diet:** Amphibians, rodents, fish, small lizards
  - **Recommended Feedings:** 1–2 weeks
- **Hog-nosed snake:**
  - **Diet:** Amphibians, rodents, fish, small lizards
  - **Recommended Feedings:** 1–2 weeks
- **Anoles:**
  - **Diet:** Insects
  - **Recommended Feedings:** Daily
- **Chameleons:**
  - **Diet:** Insects, limited greens, small lizards
  - **Recommended Feedings:** Daily
- **Geckos:**
  - **Diet:** Insects
  - **Recommended Feedings:** Daily
- **Water dragons:**
  - **Diet:** Insects, pinkies
  - **Recommended Feedings:** Daily
- **Most skinks:**
  - **Diet:** Insects
  - **Recommended Feedings:** Daily
- **Swifts:**
  - **Diet:** Insects
  - **Recommended Feedings:** Daily
- **Ameivas:**
  - **Diet:** Insects
  - **Recommended Feedings:** Daily
- **Lacertas:**
  - **Diet:** Insects
  - **Recommended Feedings:** Daily
- **Small monitors:**
  - **Diet:** Insects, pinkies
  - **Recommended Feedings:** Daily
- **Tegus:**
  - **Diet:** Insects, pinkies
  - **Recommended Feedings:** Daily

- **Day geckos:**
  - **Diet:** Invertebrates, insects
  - **Recommended Feedings:** Daily
- **Large monitors:**
  - **Diet:** Vertebrates
  - **Recommended Feedings:** Weekly
- **Large tegus:**
  - **Diet:** Vertebrates, fruit
  - **Recommended Feedings:** Weekly
- **Bearded dragons:**
  - **Diet:** Greens, invertebrates
  - **Recommended Feedings:** Daily
- **Blue-tongued skinks:**
  - **Diet:** Greens, invertebrates
  - **Recommended Feedings:** Daily
- **Uromastyx:**
  - **Diet:** Greens, invertebrates, seed
  - **Recommended Feedings:** Daily
- **Green iguana:**
  - **Diet:** Greens
  - **Recommended Feedings:** Daily
- **Prehensile-tail skink:**
  - **Diet:** Greens
  - **Recommended Feedings:** Daily
- **Softshell turtles:**
  - **Diet:** Carnivores
  - **Recommended Feedings:** Daily
- **Box turtles:**
  - **Diet:** Earthworms, fruits, veggies, crickets
  - **Recommended Feedings:** Daily
- **Wood turtles:**
  - **Diet:** Earthworms, fruits, veggies, crickets
  - **Recommended Feedings:** Daily
- **Sulcata tortoise:**
  - **Diet:** Grasses, greens, fruits
  - **Recommended Feedings:** Daily
- **Leopard tortoise:**
  - **Diet:** Grasses, greens
  - **Recommended Feedings:** Daily
- **Radiated tortoise (star):**
  - **Diet:** Grasses, greens
  - **Recommended Feedings:** Daily

Reproduction

- **Chelonians:** Oviparous
- **Monitors:** Oviparous
- **Iguanas:** Oviparous
- **Water dragons:** Oviparous
- **Geckos:** Oviparous
- **Veiled chameleon:** Oviparous
- **Panther chameleon:** Oviparous
- **Jackson chameleon:** Viviparous
- **Python snakes:** Oviparous
- **King snakes:** Oviparous
- **Milk snakes:** Oviparous
- **Rat snakes:** Oviparous
- **Corn snakes:** Oviparous
- **Most boas:** Viviparous
- **Most vipers (rattlesnake):** Viviparous
- **Colubrids—some:** Viviparous
- **Garter snake:** Viviparous
- **American water snake:** Viviparous

Zoonotic Potential

- **Bacterial**
  - Salmonella
  - Aeromonas
  - Mycobacterium
  - Campylobacter
  - Enterobacter
- **Viral**
  - None noted
- **Yeast**
  - Candida
- **Fungal**
  - None noted
- **Protozoan**
  - coccidia

Review Questions

1. Define POTZ and its importance to reptile health.

2. Why is it important to provide UV light to captive reptiles?

3. List five safety practices to reduce the potential of transmission of Salmonella.

4. What is a sign of respiratory difficulty in a reptile?

5. Define episodic breathing.

6. What are common blood collection sites in the following species?
   a. lizards
   b. snakes
   c. chelonians

7. Explain the renal portal system in reptiles.

8. What are the major differences between a reptile heart and a mammal's heart?

9. What is the major disadvantage in using a Doppler device to evaluate anesthetic depth in a reptile?

10. Describe the method of assessing anesthetic depth in a reptile and how these parameters are used to monitor recovery

11. What is dysecdysis?

12. Describe two techniques to coax a chelonian out of its shell.

13. Define autonomy.

14. Demonstrate the correct restraint of a small lizard.

15. What can be a source of injury to personnel when handling an iguana?

16. Describe how to approach a snake prior to attempting restraint.

References

Judah, Vicki, & Nuttall, Kathy, *Exotic Animal Care & Management* (2005). Clifton Park, NY: Cengage Learning.
Veterinary Clinics of North America, *Exotic Animal Practice*, January 2001, 4(1): 83–117.
http://www.fourcornersvet.com (accessed June 8, 2011)
http://www.thegentlevets.com (accessed June 8, 2011)
http://www.aphis.usda.gov (accessed January 31, 2011)

GLOSSARY

ADR: general term for cause is undetermined; stands for "ain't doing right."

alopecia: loss of hair or coat.

altrical: animals born with their eyes and ears closed and with no visible hair growth; they are entirely dependent on maternal care for survival.

analgesic: a pain-relieving medication.

anal sac: one of a pair of sacs on either side of the anal sphincter between the internal and external tissue; anal sacs contain a malodorous secretion normally expelled during defecation.

anaphylaxis: a sudden, severe allergic reaction to a foreign protein or substance; potentially life-threatening.

anesthestic: an agent that produces unconsciousness, unawareness of pain, and loss of memory.

anesthesia machine: machine that delivers gas anesthesia through a vaporizer.

Anesthesia Record: the chart used for recording all vital signs, additional drugs administered, and comments during the full period of anesthesia, from induction to recovery.

anticholinergic: a drug administered to reduce or block parasympathetic nerve impulses and reduce secretions; e.g., atropine, glycopyrrolate.

anorexia: a lack of appetite.

apnea: temporary cessation of breathing to become apneic.

anticoagulant: a chemical substance that prevents blood from clotting, coagulating.

apterium: normal, featherless tract on the skin of a bird.

arrhythmia: variation in the normal rhythm of the heart.

artifact: an image that appears on a radiograph that is not part of the animal; caused by dirt or scratches on the film or in the cassette.

ascorbic acid: vitamin C; found in citrus fruits, tomatoes, potatoes, and leafy green vegetables; necessary to prevent scurvy.

aseptic: technique used to remove debris and reduce the presence of bacteria on a surgical site.

aspiration: inhalation of a foreign substance into the lungs; withdrawal of fluids or tissue from a body cavity or node.

ataxia: lack of muscle coordination; difficult and irregular movement.

atropinase: an enzyme that counteracts the effects of atropine.

atrophy: a decrease in size or function tissues or body organs, wasting.

aural: pertaining to the ears; auditory.

auscultation: listening to sounds within the body with the use of a stethoscope.

automatic processor: machine used to develop X-ray films.

autonomy: voluntary release of the tail in reptiles; used as an escape mechanism.

avian: pertaining to birds.

axilla: armpit.

axillary: pertaining to the axilla or armpit.

bite (noun): alignment of the upper and lower teeth; medical term is *prognathism*.

blood feather: a new and growing feather nourished by a rich supply of blood.

boar: adult male of any of several mammals; a male guinea pig.

body language: body postures and expressions that communicate nonverbal messages.

bracheocephalic (brachycephalic): animal breeds with foreshortened faces, such as the Bulldog and the Persian cat.

breathing system: tube or tubes attached to the anesthesia machine to deliver anesthetic gas and oxygen to the patient. May be non-re-breathing system or re-breathing (circle) system.

buck: the adult male of any of several mammals; a male rabbit.

campylobacter: gram-negative bacteria that is often diagnosed as the causative agent for wet-tail in hamsters; potentially zoonotic.

capillary refill time (CRT): a measurement of blood perfusion.

carapace: protective shell covering some or all of the dorsal part of an animal; top half of a turtle or tortoise shell.

cat bag: zippered nylon bag designed to restrain a cat.

catchpole: long-handled snare device with a locking spring; designed to capture and restrain aggressive animals.

caudal: toward the tail; opposite of cranial.

cavy: any of various South American tail-less rodents in the family Cavidae, which includes the guinea pig; the less familiar but more correct term for guinea pig.

cecotrophs: soft feces containing nutrients, encased in a mucous membrane, and normally consumed by the rabbit; also called *night feces*.

cecum: first part of the large intestine, forming a dilated pouch in which fermentation of roughage occurs due to bacterial activity.

cephalic: vein located on the cranial aspect of the foreleg.

cere: fleshy bond between the head and the upper beak of a bird.

cerumen: waxy substance produced in the external ear; ear wax.

cerumenolytic agent: a solution used to break up and dissolve the ear wax.

chelonian: class of reptiles that includes the turtles and tortoises.

chemical restraint: administration of drugs to calm or sedate a fractious animal.

Chlamydiosis: disease caused by C. psittaci, a bacteria that infects birds and various other species, including humans; also called psittacosis.

Chlamydophila: a type of pneumonia (in humans) caused by C. psittaci.

choana: the opening between the nasal and oral cavities in birds.

choke: an esophageal blockage, characterized by dry-heaves, retching, excessive salivation, and difficulty breathing; in chinchillas, choke often is caused by ingesting pieces of food or a hairball.

clipper burn: caused by electric clippers becoming too hot and damaging the skin; may appear as a rash or small areas of nicks and cuts from the blade.

cloaca: terminal collecting sac for the digestive system, into which the urinary tract and lower bowel empty prior to waste being expelled; also, a major organ of the reproductive tract in birds and reptiles where sperm are deposited and eggs are delivered.

Coccidia: a group of protozoan parasites that infest dogs and cats.

cognitive: the state of being aware, of having a mental process that includes perception, reasoning, and judgment.

collimator: the mechanism or device on an X-ray machine used to control the dimension and direction of the X-ray beams.

computed radiography (CR): radiographic technology that uses a computer program in conjunction with specialized X-rays screens containing a reading device. Images are transmitted directly to a computer screen for viewing.

congenital: condition present at birth.

contrast: the degree or density of shades of gray, black, and white on a developed radiograph.

coprophagic: the eating of feces; normal behavior in many species, especially young animals.

cranial: toward the head; opposite of caudal.

crepuscular: animals that are more active during dawn or dusk.

Cryptosporidia: a species of coccidia a parasite found in the intestinal tract of many animals and in contaminated water.

debride: to clean and remove dead tissue from an open, usually infected wound.

deciduous teeth: first teeth ("baby teeth"), which are shed and replaced by permanent teeth.

decubitis ulcer: a pressure sore causing skin and underlying tissue damage.

deglove: complete peeling away of skin from the underlying tissue.

dewclaw: the first digit of cats, dogs, and rabbits, located on the inner side of the forelegs and, depending on dog breed, also found on the inner side of the rear legs. Dewclaws are non-weight-bearing and frequently are removed in dogs to prevent tearing injuries. Some breeds, such as the Great Pyrenees, have double rear dewclaws, and these usually are not removed as required by the AKC breed standard.

dewlap: a fold of loose skin under the chin of some species, for example, a doe rabbit.

diastolic: the period of time when the heart relaxes between beats; opposite systolic, the period of contraction.

dimorphic: species that exhibit visual differences between the sexes.

dip: an insecticidal solution poured over an animal's coat to remove external parasites.

direct radiography (DR): method of computerized radiographic technology where the images are transmitted directly to the computer and the image is not captured and scanned by a chip in a specialized plate X-ray plate. Similar to computed radiography.

diurnal: describes animals that are active during the day; opposite of nocturnal.

doe: the female of several species, such as a doe rabbit.

dorsal: pertaining to the back.

dorsal/ventral (D/V): radiographic positioning so that the primary X-ray beam penetrates from the back (dorsal) through to the belly (ventral).

dosimitry badge: device worn by X-ray personnel to record and measure the amount of radiation exposure.

Dremel®: hand-held power tool with various attachments— cutting blades and burrs. Also called hand-held rotary tool if not manufactured by Dremel®.

drilling: a low, almost guttural and rapid d-r-r-r-r sound produced by the guinea pig; used as a warning or an alert to other guinea pigs.

dry-heaves: non-productive attempts to vomit.

dysecdysis: difficulty in shedding the old, dead skin (in a reptile).

dyspnea: difficult or labored breathing.

ecdysis: the process of shedding the old skin in a healthy reptile.

ectotherm: an animal that cannot produce and regulate its body temperature and is dependant on the environment for body heat necessary for metabolic activity and health.

eczema: acute or chronic inflammation of the skin causing redness, itching, and produces lesions that become encrusted and scaly.

electrolytes: chemicals that dissolve in water and disassociate into electrically charged particles called cations or anions. Positively charged particles (cations) in body fluids are sodium (Na+), potassium (K), calcium (Ca+), and magnesium (Mg+). Negatively charged particles (anions) are chloride (Cl) bicarbonate (HCO^3), and phosphate (PO_4).

endotoxemia: having endotoxins in the blood.

Endotoxin: a bacterium that produces a toxin.

endotrachael tube (ET): a tube placed within the trachea to deliver anesthetic gas directly to the lungs and to provide a means of manual ventilation if the patient becomes apneic.

episodic: used to describe an irregular breathing pattern as occuring with general anesthesia in some species, eg. birds, reptiles.

esophageal stethoscope: a stethoscope modified with the use of a red rubber catheter for placement into the esophagus; used to monitor heart rate during surgery.

estivate: to enter a state of dormancy during hot, dry periods.

exophthalmia: protrusion of the eyeball.

exudate: fluid leaked from blood vessels, found on the skin or in affected tissue.

femoral (vein, artery, pulse): pertaining to blood vessels located in close proximity to the femur bone or the thigh.

film cassettes: specially designed cassettes with intensifying screens used to hold X-ray film.

fistula: an abnormal passage formed within body tissues.

fixed formula: commercially prepared diet in which the ingredients and percentage of ingredients do not vary with commodity prices.

flashing: the ability of birds to willfully control the size of the pupil of the eye; can occur during periods of excitement or alarm.

flow meter: the part of the anesthesia machine calibrated as a percentage to deliver a precise amount of anesthetic gas.

flow rate: a measurement as a percentage of the rate of flow of an anesthetic gas.

fluid therapy: treatment administered to restore and maintain normal body fluids.

flush valve: component of an anesthesia machine used to temporarily bypass the vaporizer and deliver oxygen to the breathing system of the anesthesia machine.

fomite: any inanimate object that carries and transmits disease causing organisms, dishes, brushes, shoes, clothing, hair, etc.

free choice: allowing an animal to eat as much as it wants and when it wants; no predetermined amount.

fur slip: voluntary release of fur as a method of escape, especially common with chinchillas.

gaping: holding the mouth open in an attempt to breathe.

gavage tube: a tube used to administer medications or supplementary feeding directly into the crop of a bird or stomach of other species.

Giardia: protozoan parasite in the intestines of most animals; causes diarrhea and abdominal pain.

gib: a neutered male ferret.

gingivitis: inflammation of the gingival tissue; inflammation of the gums.

gut-loading: feeding specific foods to prey items such as crickets so the insect is more nutritious when fed to reptiles.

gut-motility: the normal peristalsis (movement) of the intestines.

hand-fed: refers to bird chicks removed from the nest and fed a special formula by a human.

Harderian gland: a small gland behind the eye in many species, especially active during stress and illness in rodents.

heaves: a condition that occurs when animals have a respiratory compromise and use the abdominal muscles to assist in forcing out exhaled breaths.

hematoma: localized collection of pooled blood.

hemipenes: paired organs that open into the cloaca in snakes and some lizards; used together in reproduction.

hibernation: a state of dormancy during the winter, similar to estivate.

hob: a male ferret.

homeostasis: a state of balance within all body systems.

hutch: an outdoor enclosure for a rabbit.

hutch burn: a condition of the skin and urogenital area that occurs when a rabbit sits in a soiled cage.

hydration status: a measurement of hydration; percentage of dehydration in an animal.

hypertonic: an increase in tension (tonicity); a substance (fluid) that affects body cells; opposite of hypotonic.

hypoglycemia: an abnormally low level of blood glucose (blood sugar) in the body.

hypotonic: a decrease in bodily tension (tonicity), caused by a substance (fluid) that affects body cells.

iatrogenic: a hospital-acquired disease or condition as a direct result of inappropriate or careless treatment. *See also* nosocomial infection.

ilitis: inflammation of the ilium, the distal portion of the small intestine.

induction (anesthesia): the introduction or beginning period of anesthesia.

inguinal: pertaining to the groin area.

inhalation anesthesia: a volatile gas anesthetic agent that is breathed in by the patient.

integument: pertaining to the skin.

intensifying screen: a screen that is a component part of X-ray film cassettes used to produce and enhance the radiographic image.

intercostal: space between the ribs.

interstitial: cellular space within or between body tissues.

intracellular: within the cell.

intradermal (ID): between the layers of skin.

intramedullary: pertaining to the marrow cavity of bone.

intramuscular (IM): within the muscle.

intranasal (IN): to administer a drug or vaccinine through the nares (nostrils).

intraperitoneal (IP): pertaining to the area within the peritoneum, the membrane lining the walls of the abdomen and pelvic organs.

intravascular: occurring within a blood vessel.

isotonic: a solution or condition of being the same, unchanged from the normal concentration of body fluids and cells.

jill: a female ferret.

jugular/jugular vein and artery: the large vein or artery in the lateral neck region.

keel: the breast bone of a bird.

keratin: the hard tissue that forms hair, nails, and beaks.

kindle: the process of a rabbit giving birth.

kit: a young ferret.

Kurloff bodies: inclusions sometimes found in monocytes and leukocytes of guinea pigs; the function is unknown.

kvp: the high-voltage circuit on an X-ray machine.

laryngoscope: a device with a light and a tongue blade used to visualize the larynx.

lateral saphenous: describes the vein located on the lateral side of an animal's rear leg.

local anesthesia: an injectable or topical agent administered to a small, specific area to block nerves and temporarily block pain receptors.

macrodrip: a fluid deliver set that produces large drops of a solution (see microdrip).

maintenance anesthesia: the period of time during surgery when the anesthesia depth is maintained and controlled throughout the procedure

malocclusion: misalignment of the teeth; a "bad bite."

mAs: the low-voltage circuit on an X-ray machine (milliamperage and time).

mats: tightly woven, tangled masses of hair that may or may not adhere to the skin.

microchip: a small implanted device under the skin of an animal to provide permanent identification.

microdrip: a fluid delivery system that produces small drops of a solution (see macrodrip).

molting: the period of time when old feathers are shed and replaced with a new growth of feathers.

monogamous: referring to species which mate for life or maintaining only one breeding partner.

musculoskeletal: referring to the muscles and skeletal systems.

muzzle: a device used to prevent an animal from opening its mouth to bite; the area of the nose and mouth.

nebulizer: a device used to create a mist or aerosol as a method of delivering inhalation therapy.

nocturnal: active at night.

nosocomial infection: a disease that is aquired during hospitalization though passive exposure to pathogens, for example kennel cough. *See also* iatrogenic infection.

NPO: abbreviation for *nil per os*, nothing by mouth.

occlude: to close together or block off, occlude a blood vessel to temporarily block the flow of blood.

ocular: pertaining to the eyes.

open-ended question: a question that requires a response more than yes or no.

oral: having to do with the mouth; giving something by mouth.

ossify: turn to bone.

osteomyelitis: inflammation of the bone.

otitis externa: inflammation of the external ear.

otitis interna: an inflammation of the inner ear.

otoscope: instrument used to inspect the inner ear.

overbite: a condition where the upper teeth overlap the lower teeth.

overhydrate: to administer an excessive amount of fluids.

palpate: to feel with the hands.

palpebral reflex: the reflex that causes an animal to blink when the eyelids are touched.

papillae: small finger-like projections of tissue, visible at the entrance of the choana in birds.

papilloma: a wart or benign tumor arising from the skin.

paraphrase: to restate a concept or statement using different words without losing the meaning of the original.

passerine: small perching birds, including finches and canaries.

Pasteurella: gram-negative rod-shaped bacteria.

peg-teeth: the small secondary incisors found in rabbits.

pelts: the processed skin and hide of an animal (e.g., chinchilla pelts).

peristalsis: normal movement of the gut to propel ingested food through the digestive system.

phlebitis: inflammation of the vein.

photoperiod: refers to the number of hours of light and darkness.

piloerection: erection of body hair.

pinna: cartilage of the outer ear.

plastron: lower half of the shell of a chelonian (turtle or tortoise).

pododermatitis: inflammation of the skin on the plantar surface of the foot.

polyestrous: refers to species that have two or more breeding cycles during a breeding season.

popliteal: relating to the area behind the kneecap.

pop-off valve: the pressure release valve on an anesthesia machine.

porphyrin: the red pigmented fluid produced by the Harderain gland.

POTZ: abbreviation for preferred optimal temperature zone.

precocial: animals that are born fully furred or feathered, with open eyes and ears, erupted teeth, and active at birth (opposite of altrical).

pre-anesthetic: a drug or combination of drugs administered to a patient prior to induction of general anesthesia.

preening: self-grooming of feathers.

presenting complaint: the main reason or problem for which a patient is presented to a clinic or hospital.

pressure gauge: measures the amount of oxygen in a pressurized oxygen tank.

prognathism: an abnormal protrusion of the lower jaw.

proliferative ileitis: chronic inflammation of the ilium caused by a campylobacter; the condition of "wet tail" in hamsters.

Psittacine: a member of the parrot family.

Psittacine Beak and Feather Disease (PBFD): an infectious, highly contagious viral disease of birds causing high mortality.

Psittacosis: the disease in humans caused by a Chlamydia bacteria transmitted by infected birds.

pubic symphysis: the cartilaginous pelvic midline.

pulse deficit: abnormal condition where the pulse does not coincide with the heart beat.

pup: the young of many species, such as a dog pup, a rat pup.

radiology: the science of obtaining radiographs, X-rays also refer to the room or department.

radiology log: the legally required record kept in the radiology room that records all X-rays taken and all information regarding X-ray machine settings, and client and patient data.

radiolucent: refers to objects that appear dark or black on a developed X-ray film.

radiopaque: objects that appear white or very bright on a developed X-ray film.

recovery (anesthesia): the period of time when anesthesia ceases and the patient recovers from anesthesia.

reminges: the primary flight feathers of a bird.

renal portal system: refers to kidney filtration system.

rhinitis: inflammation of the mucous membranes of the nose.

ringworm: a superficial skin infection caused by a fungus; contagious and zoonotic.

rostrum: the snout or most anterior portion of the head.

Salmonella: various species of highly infectious gram-negative bacteria; zoonotic and may cause septicemia and chronic enteritis in humans.

salmonellosis: the condition of being infected with Salmonella.

scatter radiation: created when the primary beam generated by the X-ray machine stikes a solid surface and "scatters" in all directions.

scrub (noun): soapy disinfectant used to clean a surgical site; verb: the process of scrubbing the patient.

scruff (noun): loose skin at the base of an animal's neck; verb: to restrain an animal by grasping the area of skin at the base of an animal's neck.

scurvy: disease caused by lack of vitamin C.

sebaceous: pertaining to secreting glands of the skin.

sedative: a drug that induces a state of calm to reduce anxiety and excitement.

signalment: the details of an animal: species, breed, age, reproductive status, color/marking.

slip leash: soft nylon or rope with a slip noose at one end; used to assist in the restraint of a patient.

Snuffles: commonly used term to describe a disease in rabbits cause by a Pasteurella infection.

soft bills: variety of fruit-eating passerine birds.

sow: a female guinea pig.

spirochete: a coiled bacterium.

sprite: a spayed ferret.

sterile: an absence of microbes.

sternal recumbency: an animal placed down on the sternum, on the chest.

stress bars: black lines that develop on the feathers of ill birds.

stridor: a harsh sound produced in the respiratory tract during inspiration or with an obstruction in the larynx.

stylet: a fine yet fairly rigid wire or length of plastic inserted into the ET tube to guide and assist placement of the ET tube and withdrawn immediately when intubation is achieved.

subcutaneous (SQ): under the skin.

submandibular: under the lower jaw.

substrate: any material used as bedding or on the floor of animal enclosure.

surgical release form: a legal document that must be signed by the client prior to being admitted for a surgical procedure. A consent form.

systolic: the contraction of the heart, the heart beat.

technique chart: chart that determines approximate settings of the X-ray machine based on centimeter measurement of the body mass.

thorax: area of the chest between the neck and the abdomen.

tonicity: the tone or tension of body cells determined by osmotic pressure.

torticollis: also called wryneck, a condition caused by contraction of the cervical muscles creating a twisting of the head.

toxoplasmosis: a zoonotic parasite found in the intestine of all cats, especially the domestic cat.

tranquilizer: a drug administered to calm or quiet a patient.

treatment board: a whiteboard used to track patients admitted and the treatments and procedures required.

tympanic membrane: ear drum.

urethral cone: describes the external genitalia of both male and female chinchillas.

urine scald: a skin condition caused by urine soiled bedding.

urogenital: refers to the visual distance between anus and the genitals.

urolith: a calculus or stone within the urinary tract.

vaporizer: part of the anesthesia machine that converts a volatile liquid anesthesia agent into a gas vapor.

venipuncture: entering or puncturing a blood vein for the purpose of obtaining a blood sample.

vent: external opening of the cloaca in birds and reptiles.

ventriculus: the true stomach of a bird.

wet-tail: proliferative ileitis, a chronic inflammation of the ilium of hamsters, caused by a campylobacter.

Whitten effect: the synchronization of breeding females when a new male is introduced into the colony or herd.

wryneck: common name for torticollis, a condition caused by contraction of the cervical muscles, which creates twisting of the head.

Zoonotic: any disease that is transmitted directly from animals to man.

INDEX

Note: Page numbers followed by *f* and *t* represent figures and tables respectively.